STRUCTURING EXPLORATION DEALS

Volume One

STRUCTURING EXPLORATION DEALS

Volume One

By
Lewis G. Mosburg, Jr.

ENERGY TEXTBOOKS INTERNATIONAL, INC.

Structuring Exploration Deals

First Printing: May, 1983
Second Printing: March, 1985

Library of Congress Catalog Card Number: 82-082370

ISBN Number: 0-910649-02-2

Printed at IED Press, Inc., Oklahoma City, Oklahoma, United States of America.

Author's Dedication

To Robert G. Woodard.

ABOUT THE AUTHOR

Lewis G. Mosburg, Jr. is one of America's most learned writers and speakers on oil and gas, real estate, and tax-sheltered investments. Senior partner in the Oklahoma City law firm Mosburg, Sears and Kunzman, Inc., Mosburg received his LL.B degree (with distinction) from the University of Oklahoma in 1956. A member of Phi Beta Kappa and Order of the Coif while at OU, he spent the next two years in the legal department of Stanolind Oil Company (now Amoco Production Company) before entering the private practice of law in 1958.

Mosburg is a frequent lecturer for such organizations as the Institute for Energy Development, the University of Tulsa, and the Permian Basin Graduate Center. He also presents in-house seminars for Amoco Production Company, Arthur Andersen & Co., and other private business firms.

His efforts as a writer and editor have received nationwide critical acclaim. His *Landman's Handbook on Petroleum Land Titles* (IED: 1976) is now in its fourth printing and is known as the "bible" of the industry. Mosburg's other writing or editing credits for IED include the *Landman's Handbook on Basic Land Management, Contracts Used in Oil and Gas Operations, Petroleum Land Practices,* and *Problems and Pitfalls in Exploration Agreements.* In the field of tax-sheltered investments, he is the author of the popular *Tax Shelter Desk Book* (IBP: 1978) and the innovative *Tax Shelter Coloring Book* (Simon & Schuster: 1979).

Recent ETI publications by Mosburg are *Understanding Tax-Sheltered Investments* (March, 1983) and *The Economics of Oil and Gas Investment* (July, 1982).

SUMMARY OF CONTENTS

ILLUSTRATIONS

TABLE OF CONTENTS

Chapter Two
SUPPORT AGREEMENTS

Chapter Three
FARMOUT AGREEMENTS

Introduction

THE WHYS AND HOWS OF STRUCTURING EXPLORATION DEALS

Oil and gas exploration and development is increasingly becoming a joint effort. As drilling costs increase, as wells go deeper and deeper, and as new drilling frontiers are explored, more and more companies are finding it desirable to join with "partners" in the conduct of their drilling activities.

Today, the producer will find that he is seldom the sole owner of leases in a new area; desirable acreage will often be owned by other companies. And it is quite disheartening to bear the entire cost and risk of drilling only to prove up someone else's acreage.

Shorter lease terms, Pugh clauses, express drilling obligations, and skyrocketing lease costs also require a more rapid development of areas in which the company holds an acreage position. Yesteryear's option to defer drilling in favor of renewing one's leases is fast disappearing; "top leasing" is seeing to that. Yet current industry conditions—the oil and gas "glut" and restricted drilling budgets—make it more difficult to turn deals on a moment's notice or "save" leases by drilling with the company's own money.

Joint operations require a contracted definition of the respective rights and obligations of the parties. Often, the "plain vanilla" deals may ignore more advantageous arrangements than a little innovative thinking could provide for all parties. Furthermore, much "standard" contract wording is proving too simplistic to cover the myriad of problems that may arise in a seemingly innocent situation.

Toss changing tax rules into the pot, and new, more innovative recipes are clearly called for.

But do you really need to know anything about tax laws to structure a deal for your company? Most explorationists assume that a knowledge of tax laws is necessary only for their companies' tax or accounting departments, and for those Independents who are using tax "gimmicks" to sell their deals to gullible "tax-sheltered investors."

In fact, the tax incentives granted the oil and gas industry under the Internal Revenue Code are just as important to a major oil company as they are to an Independent (or to that mislabeled "tax-sheltered investor"). Industry economics are based on the availability of these legitimate tax benefits. If you are not familiar with these tax incentives, you may inadvertently misstructure your deal in such a way that tax advantages are unnecessarily lost, either to your company or to the other party.

Why worry about the other party? Because, if you deny him the return element which is derived from the tax benefits of oil and gas exploration and development (which may cost you little to give), he must insist on a tougher deal in other areas (which may cost you a lot).

The changes in the last few years in tax laws affecting the oil and gas industry have been staggering; and continued future change seems a certainty. Accordingly, a constant re-evaluation of traditional solutions appears a must for years to come, because of the changing pattern, and economic significance, of these tax laws.

But, forgetting taxes for the moment, why really worry all that much about the contract? Isn't it true that most oil deals can be done on a handshake, and that most legitimate oil operators will honor the *spirit* of the deal irrespective of the contract wording?

Honoring the spirit of the deal has been a tradition in the oil and gas industry. However, where contracts are poorly thought out (which usually means that the deal negotiations were similarly oversimplified), you may find that your understanding of the deal, and the other party's understanding, are totally different.

Even in the past, if a major dispute arose between two industry members concerning their rights and obligations under the contract, or if too many dollars were involved, both parties (and their legal departments or lawyers!) quickly went to the contract document itself to base their positions on a legalistic interpretation of the exact contract wording. Furthermore, in the last few years we have seen an unfortunate development in the industry: an eagerness on the part of many companies to insist on their exact "legal" rights to deprive the other party of a benefit or impose a burden that in fact (though not in law) appears quite unfair and contrary to the spirit of the deal. Maybe this resulted from the change in thinking that occurred once industry members began top leasing each other. Maybe the cause is the amount of dollars and the amount of risk involved in our current exploration. But whatever the cause, "relying on a handshake" is no longer a safe way to conduct business in the oil and gas industry.

It isn't the purpose of this book to cover all the legal and tax intricacies of contracts used in oil and gas exploration and development. It *is* the purpose of this book to explore the various contractual arrangements available to the explorationist to see which arrangements work best under which circumstances, to look at various structuring alternatives, and to learn of certain pitfalls to avoid.

This book is for *practical* application. First, it's unlikely that you or your management will have the time or the desire to spend days or weeks in negotiating an innovative new way to structure a farmout involving a 2,000 foot development well. Therefore, this book will deal primarily with more complicated arrangements—exploratory deals involving considerable acreage. Second, the book will focus on practical problems encountered in the industry today in currently-negotiated deals.

There is a marked absence of readily-available forms for contracts covering more complicated exploration arrangements. Accordingly, this series of volumes contains a number of sample contracts and alternate clauses. (See **Volume Four**.) These "samples" are just that—they barely scratch the surface—and often represent only a first stab by the author at suggested contract wording. Besides, a key purpose of this book is to encourage *creative* deal-structuring—to tailor the structure of the deal and the wording of the contract to the specifics of the particular prospect and the objectives and needs of both parties. So, use the forms as guides and points of departure only.

On to the deal structuring!

Chapter One

BASIC CONCEPTS

§ 1.01 General Principles of Contract Law and Negotiation

§ 1.01(a) What is a "Contract?" We've talked about "handshake arrangements" in oil and gas dealings (see the **Introduction** to this book). However, it is often far too dangerous to rely upon a nonenforceable handshake as the basis for undertaking an expensive and risky oil and gas project.

All businesses face this same problem. Accordingly, while many promises made in both business and personal life depend upon the integrity and ability of the other party to live up to his word, there are certain promises which are enforceable at law. These promises which a court will enforce are referred to as "contracts."

What is this dividing line which separates the "gentleman's agreement" from that enforceable promise referred to as a contract? For a promise to be enforceable as a contract, four elements must be present:

1. Offer and Acceptance
2. Consideration
3. Legality of the Subject Matter
4. Parties Capable of Contracting.

Let's explore each of these elements.

Offer and acceptance (frequently referred to as "mutual assent") requires a proposal made by one party which is sufficiently definite in its terms and which is accepted, exactly as made, by the other party. If the other party, in "accepting" the offer in fact changes its terms by adding new provisions, revising the terms of the proposal, or accepting on a conditional basis, it's not an acceptance; instead, a *counter offer* has been made.

A counter offer revokes the original offer: if the original offeree now decides he would have been better off to accept the first proposal, it is too late. However, the original offeror can now either withdraw from further negotiations, accept the counter offer exactly as made, or make a new counter offer of his own.

This process of offer and counter offer continues until the deal eventually falls apart, or one counter offer is accepted exactly as made. When this occurs, the mutual assent element of the contract has been satisfied.

Generally speaking, a proposing party has a right to revoke his offer at any time until it is accepted. While some states provide that a promise in a *written* offer to keep the offer open for a specified period of time must be honored, most states permit the withdrawal of an offer, written or oral, at any time—even though the offeror had agreed that the offeree would have a specific period of time in which to make up his mind whether or not to accept the offer.

What, then, about *options*? A mere promise to keep an offer open, or to enter into an agreement on pre-specified terms, is like any other promise; it is not legally enforceable unless it qualifies as a "contract." Most such promises do not constitute contracts, since they lack the vital element of "consideration" (see below). However, so long as the option is supported by consideration, and meets the other requirements for a contract, it will be legally enforceable.

Consideration is the requirement that each promise must be supported by a *quid pro quo*: something of value received by the person making the promise to which he would not otherwise be legally entitled. While consideration can be, and often is, in the form of a payment of money, anything of value which motivates the giving of the promise can constitute legal consideration. Accordingly, in oil and gas transactions the most common form of consideration is a *promise for a promise.* Thus, if I promise to drill a specified well at your request, in return for your promising to assign me an interest in specified leases—the classical "farmout" arrangement—your promise to assign the leases supports my promise to drill the well, and vice versa.

The final requirements of the contract—*legality of subject matter* and *parties capable of contracting*—while equally applicable to oil and gas transactions, usually do not create a problem in the typical exploration contract. So long as the contract does not involve an agreement to breach Federal, state, local or foreign law, and does not violate public policy, the "legality" requirement will be satisfied. The requirement for parties capable of contracting will be satisfied so long as the contracting parties are legal entities and are not minors, mental incompetents, or other parties not considered competent at law.

In certain states, or for certain purposes, additional requirements may exist for a valid contract (see, e.g., **§ 1.01(b)**, below, concerning the necessity for a written document). However, the elements listed above must exist in all states for any promise to be legally enforceable as a "contract."

§ 1.01(b) Necessity for a Written Document. The good news is that except to the extent required by local "Statutes of Fraud," there is no requirement that a contract be reduced to writing. The bad news is that contracts affecting "real estate or interests therein"—which would include oil and gas leases—do fall within the Statutes of Fraud of all states. Thus, the types of contracts with which we will be dealing in this book are required to be evidenced by a writing signed "by the party to be charged."

Notice that under this wording of the Statute of Frauds there is no requirement that the contract itself be reduced to writing, but only that the agreement between parties, whether written or oral, be "evidenced" by a signed writing. However, it is difficult to envision a significant exploration agreement that would not be reflected in a written instrument of some type.

Once the contract has been reduced to writing, an additional legal principle takes effect—the *Parole Evidence Rule.* Under this legal doctrine, once a contract has been reduced to writing, the provisions of that contract, as reflected in the written agreement, cannot be varied by anything "extrinsic" to the contract, so long as the written document is unambiguous. This does not mean that, if the contract is ambiguous, other writings or testimony as to oral discussions cannot be introduced in evidence to explain the ambiguity. A totally collateral subject previously discussed as a part of a larger package also isn't necessarily merged into a written contract not intended to cover that part of the package. Likewise, modifying agreements entered into *subsequent* to the date of the original agreement may modify the original contract, although, if the original contract was in writing, the modifying agreement normally is also required to be written. However, subject to these exceptions, evidence as to prior discussions, correspondence, or preliminary agreements, whether written or oral, is not admissible to vary the terms of the contract as finally reduced to writing.

§ 1.01(c) Formal Contract Versus "Letter Agreement." We all know what a "formal" contract is—a formidable-appearing document, obviously "legal" in nature, prepared double-spaced on 8½ x 14 inch ("legal") size paper, normally set off by an attractive blue binder. (This lets the lawyer charge a larger fee!) This formal contract will commence with such wording as, "This Agreement entered into this ______ day of ______ ," and will contain a liberal sprinkling of "Whereases," "Now, Therefores," "Witnessethes" and "Parties of the First, Second, and [ad infinitum] parts."

The letter agreement, on the other hand, is normally a much more informal document; it would be typed, single-spaced, on 8½ x 11 inch ("letter") size paper, will in fact be in the form of a letter addressed from one party to the contract to the other, and is more apt to refer to the parties as "You" and "Us."

But is this letter agreement as binding as a formal contract?

Some "letter agreements" are not intended to be binding

contracts, but merely statements of preliminary understanding, or memorandum of points, to make sure that both parties are on the same wavelength as to the current status of the negotiations. Other letter agreements, though intended to be binding, are not as clearly thought through or as accurately drafted as their more formal relative. (On the other hand, many "formal" contracts are similarly lacking in the thoroughness of their coverage or the preciseness of their wording. Otherwise there would be far less need for this book.)

So long as the letter agreement is intended to be just that—a binding *agreement* between the parties—and so long as it satisfies the four elements of a contract discussed in **§ 1.01(a)**, the letter agreement is just as binding as the so-called "formal" contract. As a matter of fact, in many instances it seems easier to reduce a deal negotiated between two oil companies to readable, understandable language when the "letter agreement" approach is used. And in some companies, a letter agreement approach will permit a more rapid formalizing of the contractual relationship, since company policy requires that formal contracts (but not letter agreements) be reviewed and approved by the legal department. (Don't ask the whys of such a policy: since both the letter agreement and the formal contract are equally binding on the parties, a requirement for legal review of one, but not of the other, without regard to the seriousness of the contract subject matter, makes no logical sense.)

§ 1.01(d) General Principles of Contract Preparation. Certain general rules will apply to all contracts discussed in this book. While certain of these considerations which extend to numerous contractual arrangements (see **Chapter Seven,** dealing with Areas of Mutual Interest) are discussed elsewhere in this book, this section will review certain principles applicable to all exploration contracts.

A critical determination in both contract negotiation and the formal preparation of the contract is whether the various "deal points" arrived at between the parties are *covenants*—agreements which the promisor is contractually obligated to perform—or mere *conditions*—provisions which must be complied with by a party to prevent his loss of certain rights, but which the party is not contractually obligated to comply with.

The dividing line between the convenant and the condition is of critical importance to the parties to an exploration deal. Assume that you own certain leases which are due to expire in six months. To secure information necessary for you to determine whether or not to drill on your leases, you have agreed to farmout certain of those leases to another operator in return for his drilling a specified well. The well is far short of total depth, appears to be poorly located on the structure, and has been subject to significant cost overruns which show no signs of abating.

If the foregoing arrangement involves a *condition*, the Farmee is now free to "walk away" from the deal without drilling the well to total depth. While the Farmee will lose his rights to an assignment of the leases, this may be a wiser decision, under the circumstances, than continuing the drilling. If, however, the arrangement involved a *covenant* on the part of the Farmee to drill the well to total depth, he is contractually obligated to continue drilling, which will give you the information you need concerning drilling on the leases retained by you.

Of course, the Farmee could always ignore his contractual obligation, cease the drilling, and hope that you would not elect to sue for damages. However, most companies are reluctant to intentionally breach their contractual obligation.

Key indications of a covenant are the words "agrees," "promises," "shall," and "will." Words indicative of a condition include "may," "if," and "unless."

Often, a Farmout Agreement, Operating Agreement, or Seismic Option will envision additional arrangements to implement the base provisions of the contract. For instance, if the Farmee performs, the Farmout Agreement must be implemented by the execution of an Assignment of the leases earned (see **§ 3.01(b)[5][a] and 3.05(c)**), and must frequently be further implemented by the addition of a

Joint Operating Agreement (see **Chapter Four**), under which the parties will develop the acreage, which, following the Assignment, is now jointly owned. There may be need for an agreement as to methods of "balancing" disproportionate "takes" of gas in "split-stream connection" situations. (See **Chapter Six**). And the contract will often provide for subsequent agreement between the parties as to the location of additional drilling sites or the nature and extent of the tests to be performed by an earning party.

A key question, thus, is whether or not to attempt to agree in advance on the form of implementing documents, such as the form of Assignment, Operating Agreement, or Gas Balancing Agreement, and to further agree, in advance, on the location of each earning well to be drilled under the contract, the exact nature of all tests to be performed under the agreement, etc.

First, let's face the fact that if we attempted in our written contract to cover every possible area of disagreement between the parties, our document would be so lengthy, and our negotiations so prolonged, it would be difficult to ever complete the negotiations and finalize the written contract. This does not mean that areas which have been left to be worked out in the future may not create serious disputes between the parties. But practicality dictates that not every potential dispute area can be resolved in advance. Likewise, where it is currently unknown exactly what future circumstances could influence such subsequent decisions, attempting to resolve potential disputes which would not arise until sometime in the future, when further information would be available, may lead to a resolution far less satisfactory than if agreement on these points had been deferred.

All this adds up to the fact that we are walking a tight rope. Leaving key items for future resolution is a dangerous practice. And there is no excuse for not covering those areas in the contract which experience has taught are key areas of negotiation. On the other hand, certain items may better be left to future resolution, despite the admitted risks of such a practice.

How can we solve some of these problems? Often, a contract will provide that Assignments will be by a "mutually acceptable" instrument, that future wells will be drilled under a "mutually acceptable" Joint Operating Agreement, or that additional earning wells will be located at a "mutually acceptable" location. As an "agreement to agree," such provisions are unenforceable if the parties cannot subsequently agree on the form of Assignment or Operating Agreement, or on the location for that additional earning well. Similarly, a reference to forms or practices "customary in the industry" can lead to serious disputes as to what *is* industry "custom," or may lock one party into a form of agreement or practice to which he never would have agreed if he had known how the matter was to be resolved.

From the standpoint of certainty, it is far better to attach the form of Assignment or Joint Operating Agreement as an exhibit to the base contract, or to provide for one party to propose a well location or testing program subject to the consent of the other party, "which consent shall not be unreasonably withheld." However, in certain instances the desire for legal certainty may have to give way to more practical considerations. For instance, if rapid action is required, timing considerations may require settling for less than an ideal document. Thus, when leases are terminating, and the preparation of a "formal" contract or reaching agreement on Operating Agreement terms or the terms of similar implementing documents to be attached to the contract would require multi-department approval or preparation time that would be disastrous, you may find it necessary to go with a document which is admittedly incomplete or less than perfect. Under the circumstances, the known risks of loss of leases may be less severe than the risks of going forward before full agreement has been established.

(Don't forget, though, that these risks are very real. Failing to reach agreement on key provisions of the deal prior to the commencement of drilling operations may lead to subsequent controversies that are far more serious and expensive to your company than the loss of one or two leases out of an extensive lease block.)

Many topics on which agreement needs to be reached will prove difficult to resolve in advance of encountering the subsequent occurrences which will trigger the obligation. A perfect example is the degree of testing. A common solution is to provide that the sufficiency of the testing, or the selection of a location, will be left to the discretion of one party to the agreement.

Before you agree to let one party's unilateral decision resolve a serious area of potential dispute, it is critical that you "know" the other party. There are certain companies where little practical difficulty may be encountered in providing that you will "test to the satisfaction of the Assignor," or accept an earning well drilled at "a location selected by the Farmee." However, even when dealing with a party with whom you have had satisfactory relationships in the past, such a provision for unilateral resolution can create a nightmare if company personnel change or too many dollars are at stake. (For suggested alternative provisions, see § § **2.04(b)[1]** *et seq.* and **3.04(b)[2]**.)

Many of the agreements which you enter into will grant an *option* or an *election* to one or the other of the parties. No matter what type of agreement you are entering into, whenever such an option or election is granted, the agreement must cover several points:

1. *How* the other party is *put on notice* of the need to *exercise* his election.
2. The *time* within which the election or option must be *exercised.*
3. *How* the option or election is to be *exercised.*
4. The *effect* of *failure to respond.*

Certain final words of caution. Sometimes you will find it necessary to structure a deal on a "panic" or "salvage" basis. In those instances, you must simply make the best deal you can get, and thus will often agree to terms which you might otherwise refuse. In some instances, you may elect to let leases expire rather than establish a reputation that you will agree to anything in an emergency. This is a "millions for defense but not one cent for tribute" type approach.

However, generally speaking, the structuring of a deal must always take into account the circumstances of the parties.

Anyone who says that any deal which they ask for they would also take probably lies about other things as well. Each company has one set of provisions that they like to see in a Farmout Agreement when they are the Farmor, and a totally different set that they prefer when they are the Farmee. There is nothing wrong with this. However, adamantly refusing to face the realities of the situation when you are in a poor bargaining situation is not good business—it's simply being pigheaded. Likewise, a key reason why so many exploration deals are poorly structured from the standpoint of both parties is the slavish adherence to "standard" deals, rather than considering structuring alternatives tailored to the specifics of this particular deal, and to the needs of both you and the other company. (See the **Introduction** to this book.) On the other hand, to be *too* greedy simply because you are in a strong bargaining position will usually come back to haunt you.

In certain circumstances, you will be considering conducting oil and gas operations in areas which are environmentally sensitive. Under those circumstances, the contract must be structured to take into account the possibility that the actual conduct of operations will be significantly delayed due to environmental action (both private and governmental); and that certain contractual provisions will not be capable of performance, either on a timely basis, or at all, due to such environmental influences. You must also consider what the impact on your company will be if these environmental considerations do significantly delay, or even prevent, the conduct of your proposed operations. (In certain circumstances, this means a determination that you would actually be better to forego operations in such areas.)

Finally, why do we even need contracts? We have touched on the fact that relying on "gentleman's agreements" can, under some circumstances, create unacceptable financial risks to your company. (See § **1.01(a)**.) However, we do not enter into contracts in the oil and gas area merely to give us the right to sue for damages (or even for specific performance) if the other party to the deal does not honor his word. In most instances, by the time such contract disputes have wound their way through the lengthy judicial process, the leases will have terminated; and the action for damages merely serves to mitigate your damages.

Why, then, the need for contracts in the oil and gas area? First, reducing your deal to a formal written contract helps insure that a true "meeting of the minds" has been reached on all significant points; and that the parties are in agreement as to what the deal actually is. Second, a party is far less apt to welsh on his word if he is firmly bound by a clear and unambiguous contract than he might be if no such clearly-worded agreement existed. Finally, if one party to the deal does go back on his word, the existence of the contract provides an avenue for at least some monetary relief to the other party.

§ 1.01(e) What You Sign is What You've Got. Far too often in oil and gas transactions, parties will scribble their signatures to a lengthy agreement without bothering to thoroughly and carefully review the terms of the document. This is a dangerous practice. Far too often someone will "read into" an agreement what he *expects to see* there, rather than what the language actually is. In addition, even when dealing with a very legitimate company you will frequently find that the contract contains "fishhook" clauses: provisions that, upon a quick reading, seem to give you just the rights and impose just the obligations to which you have agreed, but which, upon a more careful study, actually say something far different.

Remember the Parole Evidence Rule. (See § **1.01(b)**.) Once you have signed a document, you are bound by the instrument that you have signed. The fact that you did not understand the legal effect of what you are signing, or that it varies from your preliminary agreements, is immaterial: *what you signed is what you've got.* Accordingly, it is critical that you carefully review the document in its entirety to insure that your deal, as reduced to writing, is the deal that you negotiated.

The lengthier the document, the more the temptation to skim

through it, particularly as you reach what appear to be "boiler plate provisions." However, remember that what was given to you early on in the document can be taken away by a later provision. The language, "Anything in this agreement to the contrary notwithstanding," can be an introduction to a gutting of the earlier provisions of the contract. Accordingly, a thorough reading of the instrument by the person responsible for negotiating the deal before the contract is signed is an absolute must.

But can't your lawyer do this for you? Obviously, there is nothing wrong, and often much to be gained, by having an experienced attorney review any significant contractual arrangement into which you enter. But the lawyer does not know the deal that you negotiated; he merely determines that what the instrument says is legal, and that there are not any obvious pitfalls. And, a review by your legal department or outside attorney is never a substitute for a careful review of the document by whoever in your company best understands the deal.

§ 1.02 Types of Exploration Deals

§ 1.02(a) In General. There are various basic exploration alternatives available to the explorationist who determines that he wishes to jointly develop a prospect area rather than drilling his own acreage "heads up." The "whys" of reaching this basic decision *jointly to develop* are discussed in the **Introduction** to this book. However, once the joint development decision has been made, it is critical that you determine which of the basic types of exploration arrangements is both in the best interests of your company and has a reasonable likelihood of negotiation or satisfactory terms.

To make these determinations, you first must understand the deal alternatives available to you. These types of basic exploration arrangements include the following:

1. Support Agreements.
2. Farmouts and Farmins.

3. Drilling (Farmout) Options.
4. Joint Operating Agreements (Working Interest Areas).
5. Seismic Options and Exploration Contracts.

Where large areas are involved, the exploration deal negotiated often involves a combination of the above arrangements. For instance, I could agree that, if you would perform a certain amount of seismic work, you would earn the right to a Farmout of acreage owned by me over an extensive area. (See **Chapter Five** on Seismic Options.) However, the seismic alone might not be sufficient to earn the right to the Farmout. You might also have to agree to drill, in any event, not less than a specified number of wells within our area of interests. (See **Chapter Three**, concerning Farmouts.) To sweeten the deal, I might also agree to contribute a certain amount of "support" toward the drilling of those wells. (See **Chapter Two**, on Support Arrangements.) And a Working Interest Area might simultaneously be formed between us to cover the further exploration and development of our now jointly-owned acreage. (See **Chapter Four**, on Joint Operating Agreements.)

Decisions concerning structuring alternatives extend far beyond the selection of the basic type of deal which you wish to negotiate. However, this basic decision—which type of exploration arrangement should be selected—is the first step in structuring your exploration deal. To make that decision, you must have a basic understanding of the nature of each one of these types of arrangements, and what each arrangement attempts to secure for both parties to the deal.

§ 1.02(b) Support Agreements. The *Support Agreement* is a contractual arrangement under which a party with a valuable lease position agrees to contribute something of value—normally cash (though possibly acreage)—to encourage another operator to drill on his own nearby acreage. In return, the Drilling Party shares the information gained through the drilling with the party who has made the contribution.

Under the right circumstances, both parties can benefit from this arrangement. The party making the contribution will gain, at the least, information concerning his own leases, and thus has his leases at least partially evaluated at a much lower cost and risk than if he himself had drilled. This permits the Supporting Party to determine what to do with his own acreage, and whether or not to acquire additional acreage in the area. The Supported Party has his *risks* of drilling, and under certain circumstances, his *costs* of drilling significantly reduced.[1]

Support Agreements are customarily divided between *Dry Hole Contributions* ("DHCs") and *Bottom Hole Contributions* ("BHCs").

In DHCs, the contribution will be paid only if the well is drilled as specified in the contract, and is non-productive. In BHCs, the contribution is earned by the *drilling of the well,* whether dry *or* productive. Of the two, the Dry Hole Contribution is much more common.

Support Agreements are discussed in detail in **Chapter Two.**

§ 1.02(c) Farmouts, Farmins, and Drilling (Farmout) Options. The Support Agreement permits both the Supporting and the Supported Parties to develop their own acreage separately. However, in certain instances the costs, risks, and other economic factors affecting a given area may make joint development of both parties' acreage a better choice.

If joint exploration and development seems wise, one solution is to enter into a *Joint Operating Agreement.* (See **Chapter Four.**) However, if one party is more eager to drill than the other, or if there are benefits to one or the other (or both!) in having one party pay a disproportionate part of the cost of initial drilling in order to earn a greater interest in the combined acreage, a *Farmout Agreement* may

[1] Why do we say that the *costs* of the drilling are reduced only under certain circumstances? (See below and § **2.04(d)**, concerning the distinction between *Bottom Hole* and *Dry Hole Contributions.)*

be negotiated.

Under a Farmout arrangement, a party with a valuable acreage position (the "Farmor") agrees to assign an interest in those leases to a drilling party (the "Farmee") in return for the Farmee agreeing to perform certain drilling and testing obligations. The distinction between the Farmout Agreement and one variation of the Support Agreement—the *Acreage Contribution*—is that the Farmor will reserve an interest in the leases assigned.

In the Farmout arrangement, the Farmor shifts the cost and risks of drilling to another party. From the standpoint of the Farmor, he has maintained his leases in effect, obtained at least a preliminary evaluation and testing of the area, and secured an interest in production—all at little or no cost to himself. The Farmor has also "promoted" the Farmee and shifted risk, thus significantly changing the economics of further exploration and development of the prospect area.

From the standpoint of the Farmee, he has earned an interest in acreage which he desires to own and which would not otherwise have been available to him.[2]

The Farmee has also changed, in several ways, the economics of the deal to himself. First, since he will be bearing a disproportionately greater share of the costs than his interest in well revenues, the economics of drilling the initial well as a percentage of return on the investment are reduced over drilling "heads up." On the other hand, since the Farmee will now own a greater interest in the acreage, he will be able to participate in more wells, and will own a greater interest in the wells in which he does participate.

Typically, the Farmout Agreement requires the drilling of a well on acreage owned by the Farmor. However, in one variation of the

---

[2] In certain instances, other equally attractive acreage may be available in the area for purchase. The Farmee must study the economics of the deal to determine whether it is cheaper to acquire the acreage through drilling or through outright purchase.

Farmout Agreement—the *Off-Lease Farmout*—the well is drilled on acreage owned by the Farmee or some third party. (See § **3.03(b)[2].**) In other Farmout variations, the Farmee/Optionee earned the *right* to a Farmout by the drilling of a well within the area (the *Drilling [Farmout] Option),* or earns a similar right by performing specified seismic obligations (the *Seismic Option).* See **Chapter Five.**

§ 1.02(d) Joint Operating Agreements (Working Interest Areas). Even if you are quite "sold" on a prospect, budgetary limitations may make it impossible for you to accept a Farmin from other parties owning acreage in the area; or those other parties may be equally convinced of the prospect's positive economics and thus unwilling to Farmout. Or, it may prove wise to spread the costs and risks of drilling on the prospect area among several parties or to secure multi-party participation to permit a more rapid development of the prospect area. Under any of these circumstances, a *joint development* of the area is in order.

If an area is to be jointly developed, a contract must be entered into between the drilling parties to define how this joint development will be conducted. This agreement must define the contractual rights and obligations of the parties over three stages of the operation—the initial testing of the area, the area's further development, and the operation of any producing properties.

The contract which defines the rights and obligations of the parties under such an arrangement is referred to as a *Joint Operating Agreement.* This agreement must accomplish several objectives. First, it must *pool the interests* of the co-owners—from a practical, if not from a legal, standpoint. Second, it must provide a *method* for the *joint exploration, development,* and *operation,* i.e., how disagreements as to the conduct of the operations are to be resolved. It must provide *how* the *costs* and *revenues* of the joint operations are to be *shared* among the parties. And finally, the agreement must *designate an Operator* to be in charge of the day-to-day operations, and must specify the powers, duties, compensation, and replacement of this Operator.

Often, the Joint Operating Agreement covers as little as a single drilling and spacing unit, where this DSU includes leases owned by several parties. Under this agreement, all the parties owning an interest in the drilling and spacing unit would have shared in the costs and production of the well in one form or another in any event—if not by agreement, then as a result of forced pooling. However, in other instances, a much broader area is involved. Under these circumstances, it would have been, and is, possible to develop the area by each party drilling solely on his own leases. However, at other times it may be more advantageous for the parties to pool their interests throughout this larger area. It is such a multi-drilling and spacing unit pooling arrangement that we refer to when we speak of forming a *Working Interest Area.*[3]

The impact of the formation of the Working Interest Area is that the parties will share in the costs and production of the wells drilled based on their relative acreage position *throughout the Contract Area,* rather than on the ownership of the leases *within the particular drilling and spacing unit* on which a given well is drilled.

Most exploration contracts which involve the earning of acreage by a "Drilling Party" are followed by a joint development of the now jointly-owned leases. Thus, Farmout and Seismic Option arrangements are normally supplemented with Working Interest Area (Joint Operating Agreement) arrangements. Similarly, the Off-Lease Farmout (see **§ 1.02(c)** and **§ 3.03(b)[2]**) might make little sense to the Farmor if a Working Interest Unit were not also being formed.

While Support Agreements are not, as a general rule, supplemented by the formation of Working Interest Areas, in other instances a support contribution and the formation of a Working Interest Unit will be combined into a single contractual arrangement.

---

[3] The Working Interest Area is sometimes referred to as a *Working Interest Unit.*

Joint Operating Agreements and Working Interest Units are discussed in detail in **Chapter Four**.

§ 1.02(e) Seismic Options and Exploration Agreements. Often, if an area is "frontier" in nature, no one can intelligently conduct initial testing until further seismic or similar information can be secured. A prospective Farmee may be unwilling to drill a deep test without further information; however, unlike the situation in Canada, this party, if he does not own leases in the area, has no right to shoot such seismic without the consent of the leasehold owners. Similarly, if the acreage is merely "on trend," where should the wells be drilled? The leasehold owners, on the other hand, due to budgetary limitations or the like, may be unwilling or unable to expend the money for the conduct of the necessary geophysical operations.

Under these conditions, a *Seismic Option* may solve the problem for both parties. Under this type of arrangement, the Optionee/Farmee earns the right to Farmout on pre-established terms by the conduct of specified geophysical operations. Once the seismic has been performed, the Optionee is not *required* to enter into the Farmout arrangement if the geophysical data is discouraging. However, he is not faced with the problem of having expended significant amounts of money in shooting seismic only to discover that leases in the area are no longer available. From the standpoint of the Optionor, at the worst, he now has additional information concerning the value of his leases. At the best, he will also have secured a testing of the area, as well as a "partner" for further development if the initial wells are productive. (Of course, he has given up an interest in his acreage position to secure this information and/or testing.)

The Seismic Option just discussed involves an arrangement between two exploration companies, normally over a very broad "trend" area. In another variation, a landowner may own a large ranch or farm. The acreage is potentially quite valuable for oil and gas exploration and development (again, possibly "on trend");

however, without further information, the terms on which the Lessor is willing to lease may be unacceptable to potential Lessees.

Here again, a Seismic Option—in this instance from a Lessor rather than from another working interest owner—may solve the problem. The Lessor agrees that, upon the conduct of specified seismic by the Lessee, it will permit the Lessor to select a certain amount of acreage for lease on pre-agreed terms.

The Lessee is taking less risks than would have been the case if he had taken a lease on the entire acreage upon terms which the Lessor would accept. The Lessee is also permitted to "highgrade" the acreage which he actually leases. From the Lessor's standpoint, he will receive a payment for having granted the Seismic Option, even better terms for the leasing of that acreage selected by the Seismic Optionee/Lessee, and information concerning the entire block;[4] and an upgrading of the terms he can demand for the leasing of the balance of the acreage. (Of course, other lessees may be "turned off" on leasing the balance of the acreage, since it was obviously not considered desirable by the operator who now possesses seismic information.)

There are various other types of exploration contracts which can be entered into between working interest owners for the exploration of extensive "frontier" areas.

The Joint Operating Agreement under which a typical Working Interest Area is formed most frequently covers a single prospect. The broad-area Exploration Contract normally covers an area extensive enough to contain numerous potential prospect areas. While Working Interest Area agreement and the broad-area Exploration Contract must cover many of the same general subjects, the Exploration Contract is normally much more complicated than the Joint

---

[4] This is one situation in which you may be forced to turn over your *seismic interpretations, and not merely your raw data,* to the other party. See **Chapter Five**.

Operating Agreement under which a Working Interest Area is formed. For instance, the Exploration Contract may provide for an initial joint conduct of *seismic* over all, or various portions, of the Exploration Area. After this initial phase of the operation has been completed, an *exploratory phase* will arise under which test wells are drilled on the various prospects identified during the geophysical phase. The agreement will then move into a third phase covering the further *joint development* of the prospects which have proved productive.

Often, operations covering all three phases will be conducted simultaneously under the Exploration Contract. In one portion of the Exploration Area, the conduct of seismic may be taking place. In another portion, where the seismic has proved interesting, initial testing may be underway. And, in another portion, exploratory wells previously drilled may have lead to an ongoing program of additional development drilling.

Seismic Options and other types of broad-area exploratory contracts are discussed in **Chapter Five**.

§ 1.02(f) Areas of Mutual Interest. Once an area has been initially tested, all parties to the agreement will now possess information which may make it desirable to secure leases on "open" portions of the Contract Area, or to negotiate Farmouts or similar agreements with other parties owning acreage in the area. There is nothing wrong with each of the parties who are sharing the common information competing with each other for the acquisition of these leases. However, it may prove more desirable—to reduce unnecessary competition and to avoid complicating the existing contractual "sharing" arrangement between the parties—for acreage acquired by any party to the initial contract to be shared with the other parties to the agreement.

The contractual arrangement under which each party to an agreement is given a right, at his option, to share in other acreage acquisitions by another party to the agreement within a specified area, is referred to as an *Area of Mutual Interest* or "AMI" provision.

The AMI provision is not unique to any given type of contract, but may be found in Farmouts, Joint Operating Agreements, Seismic Options, and other types of Exploration Contracts. While less common, a Support Agreement could even be coupled with an Area of Mutual Interest arrangement.

Area of Mutual Interest provisions are discussed in **Chapter Seven.**

§ 1.03 Selecting the Proper Deal

§ 1.03(a) In General. There is no "school solution" as to which type of exploratory arrangement is appropriate for the exploration and development of a given prospect area. Instead, the explorationist must take into account a number of factors in determining which deal alternative makes the most sense for this particular prospect area. An understanding of the basic principles of petroleum economics (see § **1.03(c)** and its subsections) is critical in reaching this determination. Similarly, the impact of the Federal income taxation of oil and gas transactions must be taken into account (see § **1.03(b)** *et seq.*). Finally, other factors, such as the degree of risk and type of prospect, budgetary limitations, and lease expiration dates, will influence how the deal should be structured. (see § **1.03(d)**.)

As mentioned in the **Introduction**, far too many complicated exploration deals are "structured" by merely taking a standard form and attempting to force the deal to fit this form rather than tailoring the form to the deal.[5] If this were the best way to find oil, a clerk or a computer would be an equally suitable, and a much cheaper, source of deal negotiation. However, even the computer has not replaced the

---

[5] Often, these "standard deals" are based upon formats and deal terms developed years ago, and have never been changed to reflect changing industry conditions.

human brain as the best resource for structuring a mutually-profitable method of developing oil and gas reserves. Likewise, "standard" deals often do not take into account pitfalls that may be common to the exploration and development of certain, but not all prospects. Accordingly, the dedicated explorationist will force himself, and even take pride in, structuring the deal to fit the circumstances.

The object of such "tailoring" is *not* an arbitrary exercise of "innovative" structuring for structuring's sake. It is rather to make a deal where other people have failed; to make more money for your company (and often for the other party!) than would flow from the "standard" arrangement; and, above all, to protect your company against an arrangement which might prove, unnecessarily, an economic disaster.

§ 1.03(b) Impact of Federal Income Tax Laws

§ 1.03(b)(1) Why Worry About the Tax Laws? As many U.S. taxpayers have learned, to their sorrow, it's not what you *make*, but what you get to *keep—after taxes*—that counts. Petroleum economics are always computed on an after-tax basis. Accordingly, it is impossible to determine which method of structuring a deal is in the best interests of your company without understanding the impact on the arrangement of the Federal income tax laws that apply to the taxation of oil and gas transactions.

The oil and gas industry is subject to the same tax laws that apply to any other U.S. taxpayer. In addition, there are numerous special tax incentives granted to those engaged in oil and gas exploration and development, as well as certain tax *disadvantages* that are unique to oil and gas activities. Furthermore, the impact of these rules varies between corporations and individuals and between "major" oil companies and "independents," even though all of these parties are participating in the same wells.

As discussed in the **Introduction** to this book, many explorationists pride themselves on their ignorance of Federal income tax laws, thinking that tax laws are for the accounting or tax departments to worry about. However, the accounting and tax departments don't negotiate the deal. Even when a deal is submitted to your tax specialists, their analysis will be affected by the problem facing any specialist—"tunnel vision."

This book is not intended to make a tax specialist out of you. However, as an explorationist you must understand the basic Federal income tax consequences of the various deal structures available to you. A wrong tax structure can turn what would otherwise have been a "good" deal into an after-tax disaster for your company; the right structure can turn a marginal deal into a profitable one, or make a profitable deal even more profitable.

Understanding the tax consequences to your company, alone, is not enough; you must also understand the tax consequences of the arrangement to the other party. By negotiating a deal which maximizes the tax advantages to the other party, you may be able to give that party tax benefits which are of immense value to him, but will have little, if any, adverse consequences—tax or otherwise—to you. If so, you can structure an arrangement which is more favorable to you from an economic standpoint without adversely affecting the economic return to the other party. On the other hand, if you negotiate an arrangement which unnecessarily creates negative tax consequences to the other party, he must insist that your share of costs be greater, or that your interest in the oil and gas reserves and revenues be less. At best, you have forced your company into a less favorable deal than could otherwise have been negotiated. At worst, a deal which could have been worked out on a satisfactory basis to both parties now falls apart.

Taking tax consequences into account in structuring a deal does not necessarily mean maximizing the tax benefits! We are still looking at the "bottom line" to both parties—which type of arrangement will give your company (and possibly the other party as well) the *most money after taxes.* If maximizing tax benefits means giving up two dollars in economic return for a dollar of tax savings, then tax structuring must give way to more beneficial non-tax considerations.[6] On the other hand, if two dollars in tax benefits can be gained at a loss of one dollar in economic return, the tax-oriented structuring makes sense. (And if that two dollars in tax benefits can be secured at no economic cost to either party, so much the better!)

The explorationist who understands our Federal income tax laws will negotiate better deals for his company; will make deals where other parties have failed; and, most important, will avoid arrangements which may create an economic catastrophe for either, or both, parties to the deal when tax considerations are taken into account.

The reasons that so many explorationists shy away from tax considerations are two-fold. First, the Internal Revenue Code is very complicated, and is constantly changing. Second, the various provisions of the Internal Revenue Code appear highly illogical if read in a vacuum, unless there is an understanding of the social and political factors which may have led to the enactment of a given Code provision.

The lack of a logical interrelationship among the various provisions of the Internal Revenue Code is in itself logical when you understand what our tax laws are aimed at accomplishing. Obviously, a key purpose of our tax laws is to raise the money required to run the Federal government. But a second, and equally significant, purpose is to encourage or discourage various courses of conduct which changing Administrations and Congresses consider socially desirable or undesirable. Thus, as economic and social conditions change, and as conservative, moderate, and liberal Administrations and

---

[6] Hereafter, we will talk of "tax" and "economic" considerations as if they were separate concepts. In fact, the tax consequences are simply *one factor* in evaluating the economics of a petroleum project.

Congresses come and go, provisions are added to or deleted from the Internal Revenue Code. We see provisions in our tax laws encouraging oil and gas exploration and development (see the discussions of the deductibility of Intangible Drilling and Development Costs and Percentage Depletion in § § **1.03(b)(6)** and **1.03(b)(4)**), coupled with other Code provisions which tend to discourage oil and gas activities. (See the same sections, dealing with the negative treatment of "productive-well IDCs," the repeal of Percentage Depletion for "major" oil companies, and the enactment of the Crude Oil Windfall Profit Tax.)

The apparent lack of logic and consistency among the various sections of the Internal Revenue Code have a double impact on the explorationist. First, they make the Internal Revenue Code and its effect on the structuring of the deal much more difficult to understand. Second, they create a real challenge in deal structuring, since "tax accounting" does not necessarily (or even ordinarily) reflect economic reality. Thus, a deal can be structured in such a way as to create a "taxable loss" for your company's Federal income tax return, when the transaction has in fact, resulted in an economic gain. On the other hand, it is possible—when a deal is mis-structured—to have suffered economic disaster, but, for tax purposes, to be treated as if the transaction were highly profitable.

Simply stated, the tax treatment of any exploration deal is governed by the arbitrary rules contained in our tax laws. We attempt to structure a deal in such a way that a profitable transaction is treated as being less profitable from a tax standpoint (or better still, as having created a taxable loss), thus reducing the amount of dollars which your company is required to pay the Federal government in the form of income tax. On the other hand, we similarly—and with equal fervor—attempt to avoid creating *"Phantom Income"*—the creation of "income" or "gain" for tax purposes when the transaction actually resulted in an economic loss.

As mentioned in the **Introduction**, it is not the purpose of this book to make you an expert in the intricacies of the Internal Revenue

Code as it applies to oil or other areas. The explorationist does not need this; this is what your tax and/or accounting department is for. However, an understanding of the basic principles of oil and gas tax is necessary for you to know how to structure your deals; whether or not to accept a deal proposed by someone else; and to know when a trip to your tax specialist is necessary.

Consulting with your tax consultants before an intricate deal is finalized is critical. Tax laws are complicated, and they are constantly changing. (For instance, the Economic Recovery Tax Act of 1981 ("ERTA") has significantly changed a number of the tax principles on which our exploration agreements were formerly based. (See § § **1.03(b)[2]**, **1.03(b)[4]**, and **1.03(b)[5]**.)

There are two annually-updated one-volume treatises that serve as the industry "bibles" when it comes to the Federal income taxation of oil and gas transactions. These books are Burke and Bowhay, *Income Taxation of Natural Resources* (Prentice-Hall, Inc.); and Houghton (ed.), *Miller's Oil and Gas Federal Income Taxation* (Commerce Clearing House, Inc.).

§ 1.03(b)[2] How Do Our Basic Tax Laws Operate?

§ 1.03(b)[2][a] Calculating Taxes Payable. Whether you are in the oil and gas business, the manufacturing of automobiles, or work for a salary, certain basic tax principles apply to all United States taxpayers.

A U.S. taxpayer pays a tax on his *taxable income.* For corporations, this is the taxpayer's *gross income*—his income from all sources, including *gains* (but not the gross receipts) from all sales of property—less his *deductions* (his expenditures which are currently deductible for Federal income tax purposes). (See § **1.03(b)[2][c]**.)[7]

[7] The calculation of *taxable income* for individuals is slightly more complicated. For the individual, *gross income* less *business deductions* equals *adjusted gross income.* Thereafter, the individual is permitted to deduct certain specified *itemized deductions*—certain expenditures specifically made currently deductible for individuals under the Internal Revenue Code, such as alimony and interest expense—to arrive at his *taxable income.*

After determining its taxable income, the U.S. taxpayer then arrives at his *tax liability* by multiplying his various levels and types of taxable income by the applicable *tax rate.* This tax rate varies, depending upon the *type* of income—*Ordinary Income or Long Term Capital Gains—and* by income level. The tax rate will also vary based on what *type of entity* the taxpayer is—individual, corporation, trust, or estate.

Prior to the Economic Tax Recovery Act of 1981, hereinafter at times referred to as "ERTA," drilling of oil wells made more sense, tax wise, for individuals than for corporations. Due to the difference between the top corporate tax rate of 46% and the top individual tax rate of 70%, the individual would secure greater benefits from the deductibility of the Intangible Drilling and Development Costs (see § **1.03(b)[6]**) than it would for a corporation. Many Farmout Agreements were based on this tax rate distinction. However, since ERTA, individuals are now taxed at a maximum 50% Federal income tax rate. (The current distinction between the 46% corporate tax rate and the 50% individual tax rate is of relatively little economic significance.)

Tax liability does not necessarily represent a taxpayer's taxes payable. *Taxes payable* are arrived at by applying certain *tax credits* created under the Tax Code as *direct offsets* against tax liability.

The foregoing method of tax calculation utilized by the Internal Revenue Code is further illustrated in **Figure 1.**

§ 1.03(b)[2][b] Taxable Entities. We've mentioned that the *type of* taxpayer involved affects the tax rate at which taxes must be paid by that taxpayer. When multiple parties wish to participate in a deal, the type of entity they utilize to conduct a transaction also affects that transaction from a tax standpoint.

Individuals, corporations, estates, and trusts are all legal entities which are considered *taxable entities,* i.e., they must pay a tax on their taxable income, and which offset *only the taxable income of that entity* with any deductions or tax credits created by the entity's

GROSS INCOME

Less

DEDUCTIONS/BUSINESS DEDUCTIONS

Equals

TAXABLE INCOME

Times

TAX RATE

Equals

TAX LIABILITY

Less

TAX CREDITS

Equals

TAXES PAYABLE

Figure 1: Calculating Taxes Payable

activities. However, under the U.S. Internal Revenue Code, "partnerships" are not taxable entities and thus do not directly pay any tax on their taxable income.[8]

What practical difference does it make that a partnership, as defined in Subchapter K, does not constitute a taxable entity for U.S. Federal income tax purposes? This means that a multi-party arrangement, if structured as a Subchapter K activity—a partnership *for tax purposes*—will be able to take any taxable losses and any tax credits at the "partner," i.e. (the individual participant) level rather than at the transaction level. In contrast, if the parties had decided to form a "corporation" to conduct their activities, all deductions and credits would have to be taken at the transaction level by the corporation, and would not pass through to the individual participants.[9]

Thus, the partnership is referred to as a "pass through" or "flow through" vehicle for tax purposes.

Since the partnership is not a taxable entity, it also pays no tax on its income—all taxable income or taxable gain is taxed directly to the individual participants in the transaction. This, again, contrasts with the treatment which would result, for tax purposes, if the parties had decided to structure their transaction by forming a corporation; if so, the income would be taxed first at the transaction (corporate) level,

---

[8] For tax purposes a "partnership" is a multi-party activity *defined as a partnership* under Subchapter K of the Internal Revenue Code. Many arrangements which do not create partnerships for state law purposes are classified as partnerships under Subchapter K. See **§3.04(b)[2][c-5-b-2]**. On the other hand, some entities which do constitute partnerships under state law are not partnerships for purposes of our tax law. See below.

Unlike the U.S. tax laws, partnerships in Canada are taxable entities. This creates complications for U.S. corporations conducting operations in Canada, either directly or through a Canadian subsidiary, or for Canadian companies conducting operations in the United States.

[9] One type of corporation—the "Subchapter S" corporation—is taxed, in general, in the same manner as a partnership. However, due to the technical requirements for the proper formation of such entities and continuation of their Subchapter S status, and due to certain tax *disadvantages* of such arrangements, the Subchapter S format is not widely used in the conduct of oil and gas activities.

and then again at the individual "shareholder" level when the monies were distributed to the participants in the deal.

Obviously, there are certain business advantages to using the corporate structure—centralization of control and limited liability, to mention only a few. However, to avoid the "double taxation" and loss of "flow through" tax benefits, which would result if a corporate arrangement were used in structuring a multi-party transaction, most oil and gas deals are either structured utilizing a co-ownership (tenants-in-common) arrangement or a partnership.[10]

Partnerships do not escape scot-free from the impact of our Federal income tax laws. While a partnership is not a *taxable entity*, it *is* a *reporting entity* and must file an "Information Tax Return" each year with the Internal Revenue Service, setting forth the tax consequences of its transactions and how these tax consequences are allocated, under the partnership agreement, to the individual partners participating in the deal.[11] Since quite a few headaches go along with the Information Tax Return (see **§ 3.04(b)[2][c-5-b-4]**), most multi-party industry deals have involved an election **to be excluded** from the provisions of Subchapter K, so that the transaction could be taxed as a "co-ownership" situation rather than a partnership.[12]

Just as a business venture can be a partnership for tax purposes even though it remains a co-ownership for other purposes, there can be "tax corporations"—multi-party ventures which, for state law purposes, are co-ownerships or partnerships. But due to some

---

[10] The Internal Revenue Code defines "corporation" to include "Associations Taxable As Corporations." Accordingly, it is critical that exploration deals be structured in such a manner as to avoid this "Association" status. See below.

[11] The Revenue Act of 1978 imposes stiff penalties and extended audit periods for failure to file such Information Tax Returns.

[12] This almost automatic decision to avoid Tax Partnership status by electing to be excluded from Subchapter K often ignored certain very real benefits that might flow from being treated as a partnership for tax purposes. See **§ 3.04(b)[2][c-5-b]**, *et seq.*

inadvertent (and avoidable!) blunder, they are, for tax purposes, taxed as corporations. Since such "Associations Taxable As Corporations" will have all of the tax *disadvantages* of a corporation but none of the corporation's business advantages, it is critical to avoid misstructuring a multi-party deal so that it advertently becomes an "Association." (See § **3.04(c)**.)

§ 1.03(b)[2][c] Classification of Expenditures. When expenditures are made in connection with any business transaction, such expenditures will be characterized in one of three ways for Federal income tax purposes:

1. Expenditures for items which will have a *future useful life*, but which will *deteriorate in value over time* are characterized as *Amortizable Capital Expenditures.* (An example of such an expenditure is the purchase of a car.) Such expenditures are not currently deductible for Federal income tax purposes, but must be "capitalized," and become the taxpayer's "basis" in the property. The expenditure is recovered, for tax purposes: a) by a "depreciation" deduction over the estimated useful life of the property if the asset is one which will automatically depreciate in value over time due to wear and tear or obsolescence; or b) by a "depletion" deduction as the related natural resource is extracted if a "wasting" asset, which is consumed as the asset is used, is involved.

2. Expenditures for assets with a *future useful life* which *will not depreciate in value over time* are referred to as *Non-Amortizable Capital Expenditures.* (An example of such an asset is land.) Such expenditures likewise must be "capitalized" for Federal income tax purposes and become the taxpayer's "basis" in the property. However, *no amortization* of the expenditure is permitted; instead, when the property is sold, the taxpayer's *Basis is used to reduce his taxable "gain" on the sale.*

3. Expenditures for items which *do not have a future useful life* are referred to as *Expenses*. (Examples are salaries and wages.) Such expenditures are *currently deductible* for Federal income tax purposes.

Note that in each case the taxpayer, in one way or another, is entitled to offset otherwise taxable income with the full amount of his expenditure. In the case of Expense items, the offsetting is through the current deduction. In the case of the Amortizable Capital Expenditure, the offsetting is taken over the life of the property or the extraction of the natural resource. In the case of the Non-Amortizable Capital Expenditure, the offset is against the otherwise taxable gain on sale.

So, if we get the same amount of deduction in any event, why do we care *how* the expenditure is characterized? (Since most of our tax-oriented structuring is based upon insuring the proper classification for expenditures, the answer to this question is extremely important.) The answer is based on a principle referred to as the *Time Value of Money*—the principle that the longer an expenditure or other burden can be put off, the less it actually costs the paying party; and the quicker revenues or other benefits are received, the greater their value. This is because money has a "use value"—if you have money, it can be put to work for you to earn more money; if you do not have money and need it, you will be required to pay for borrowing the use of that money from another party. (See § **1.03(c)[2][a]**.)

Under the Time Value of Money principle, an expenditure which can be currently deducted for Federal income tax purposes, i.e., an Expense item, is of the greatest value for tax purposes to the paying party, since the deduction and its offset against taxable income, with the resulting reduction in taxes payable, is received at the earliest possible date. An Amortizable Capital Expenditure, while not currently deductible (and thus less valuable, tax-wise, than an Expense item), can still be amortized over the life of the property, and is thus of more value than an expenditure which cannot be used to offset taxable income until the asset is sold. This is particularly true since the amortization of the expenditure frequently need not be taken ratably over the life of the property ("Straight Line Amortization"), but can be "accelerated," so that greater deductions are taken in the earlier years, with a corresponding reduction in the amount of deductions later in life of the property.[13] Likewise, such amortization can often be taken over an artifically short *Recovery Period.* (And the shorter the Recovery Period, the quicker the amortization, with a resulting increase in its Time Value.) The Non-Amortizable Capital Expenditure is the least valuable form of expenditure for tax purposes, since the offset against taxable income cannot be taken until the property is sold, and since such offset is often against income which would be taxed only at the less onerous *Long Term Capital Gain* rate, rather than at the higher *Ordinary Income* rate. (See § **1.03(b)[2][d]**.)

§ 1.03(b)[2][d] Tax Treatment of Revenues. Insofar as the revenues from a transaction are concerned, income of all types, from whatever source it is received, as a general rule must be included within the taxpayer's *gross income.* This includes such items as salaries, interest received on monies loaned by you, and gains on the sale of property. However, in certain instances, a portion of the revenues from a transaction may be received at least partially tax free by virtue of offsetting deductions for Depreciation or Depletion.

Special rules also apply to the taxation of revenues from the sale or exchange of property. While sales and most exchanges of property are "taxable events," i.e., have present tax consequences, the method of computation of this tax differs from the tax on income received in other ways. Normally, all revenues from a transaction are subject to tax. However, when the revenue is from the sale or a taxable exchange of property, only the "gain" on the sale is subject to Federal income taxation. This "taxable gain" is based upon the *difference*

[13] The amortization, whether accelerated or straight line, is still limited to the taxpayer's basis in the property. (But see § **1.03(b)[4]** as to Percentage Depletion.)

between the *amount realized on the sale* or taxable exchange, and the taxpayer's *Adjusted Basis* in the related asset.[14]

The tax consequences of the sale, or taxable exchange of an asset, will vary, based on the type of property involved and the length of the taxpayer's "holding period." All gains on the sale or taxable exchange of property held for 12 months or less are taxed at the 46% to 50% Ordinary Income tax rate.[15] As to property other than *Ordinary Assets*—property held primarily for sale to customers, e.g., inventory or stock in trade—the gain on the sale of such property, if held for over twelve months, is taxed at the special Long Term Capital Gain rate of 28% for corporations and up to 20% for individuals.[16] However, when an Ordinary Asset is sold, the gain is taxed as Ordinary Income, irrespective of the holding period.

Clever tax strategists are constantly on the lookout to find ways of structuring transactions to avoid or reduce tax liability. The combining of Depreciation Deductions (particularly deductions for

---

14 A taxpayer's Adjusted Basis is his original capitalized cost in the property, increased by the amount of any related capital expenditures and reduced by the amount of any amortization of his Basis, such as deductions for Depreciation or Depletion.

If the sale results in a taxable loss, i.e., the taxpayer's Adjusted Basis in the property is greater than the amount realized on the sale, all or a portion of that loss may be deductible either currently or over time. However, in most sales of developed oil and gas properties, the high amount of taxable deductions incurred in connection with the development will leave the taxpayer with a relatively small Adjusted Basis in the property.

15 While tax rates for both corporations and individuals are graduated, as indicated in § **1.03(b)[2][a]**, it would be highly unusual for a corporation or individual engaged in the oil and gas search not to be in the highest Federal income tax bracket.

16 The Long Term Capital Gain rate is fixed at 28% for corporations. For individuals, 60% of the Long Term Capital Gain is excluded from any taxation, and 40% is reported as part of the taxpayer's Ordinary Income. Since the maximum Ordinary Income tax rate for individuals, as a result of ERTA, is not 50%, and since most individual taxpayers engaged in oil and gas transactions will fall in this highest bracket, the *maximum* Long Term Capital Gain rate for individuals is 20%.

A taxpayer's assets subject to Long Term Capital Gain treatment include Capital Assets: property held for investment or business use; and "Section 1231" Assets: depreciable business property and business real estate. While the tax treatment for taxable *losses* differs for Capital Assets and Section 1231 Assets, the treatment of any Long Term Capital Gain on sale is the same.

Accelerated Depreciation), which will offset taxes payable at Ordinary Income tax rates, and utilization of the Investment Tax Credit (see § **1.03(b)[5]**) with the reduced Long Term Capital Gain rates payable when property is sold after a 12-month holding period, would permit a taxpayer to purchase assets of questionable economic potential merely to benefit from *Capital Gains Conversion*. Under this tax-planning strategy, if the asset could merely be sold for an amount equal to the taxpayer's original investment in the property, the lower tax rate on the Long Term Capital Gain versus the previous Accelerated Depreciation Deductions would cause the taxpayer to benefit merely because of this difference in rates. (The Investment Tax Credit would provide more tax benefits.)

To avoid such game playing, Congress in certain specific instances has provided that certain gains on sale must be "Recaptured" at Ordinary Income (not Long Term Capital Gain) rates, even where the property is a Capital or Section 1231 Asset which has been held for over 12 months. Similarly, when property has been held for less than a specified period of time, previously-claimed Investment Tax Credit benefits may also be subject to recapture.

This is not to say that recapture eliminates the positive nature of the Accelerated Depreciation and Investment Tax Credit benefits; since the taxpayer has had the use of the money saved through these tax benefits during the period of time that he owned the asset, Time Value principles decree that he still has come out ahead. However, recapture at Ordinary Income tax rates, where applicable, does reduce the advantages that otherwise would have been secured through Long Term Capital Gains Conversion.

Generally speaking, the *exchange* of property for property is treated in the same manner as the sale of such property. Obviously, one complicating factor is that the *amount realized on the sale* is not as easy to arrive at as if that amount had been a sum of cash. Accordingly, our tax laws provide that the *amount realized* on a taxable exchange of property will be the Fair Market Value of the property received as of the date of exchange.

In certain instances, the Internal Revenue Code permits an exchange to be made on a "tax-free," i.e., non-taxable, basis. Such *non-taxable exchanges* include:

1. The exchange of investment or business property for investment or business property of a like kind, e.g., real estate for real estate, or oil-field equipment for oil-field equipment. (The "Like-Kind Exchange.")
2. The sale by a corporation of its stock, or the exchanges of its stock for property.
3. The contribution of property or cash to a partnership in return for an interest in that partnership.

The taxpayer does not permanently eliminate the tax on the "non-recognized" gain when a tax-free exchange is made. Rather than receiving a "step-up" in his Basis in the newly-received property to its current Fair Market Value as of the date of the exchange, as would have been the case if the exchange had been a taxable one, the taxpayer is required to "carry over" his Basis in the property exchanged, which now becomes his Basis in the newly-received property. However, since he has at least deferred the taxation of the gain, Time Value principles make the non-taxable nature of the transaction quite valuable to the transferring parties. The tax effects of disposing of oil and gas properties will be discussed further in **§ 1.03(b)[7]**.

§ 1.03(b)[3] Taxation of Oil and Gas Transactions—An Overview. As discussed in **§ 1.03(b)[1]**, no one expects you, as an explorationist, to become an expert in the details of the Federal income tax laws as they apply to the taxation of oil and gas transactions.[17] However, in addition to a knowledge of the general principles of Federal income taxation discussed in **§ 1.03(b)[2]**, it is necessary that you understand certain of the provisions of the Internal Revenue Code which uniquely apply to oil and gas transactions.

---

[17] See **§ 1.03(b)[1]** as to additional reading sources on the details of oil and gas taxation.

As mentioned in § **1.03(b)[1]**, there are a number of such special "oil" provisions in the Internal Revenue Code. Certain of these provisions make oil and gas exploration and development more attractive; others reduce the attractiveness of the oil and gas search. A knowledge of both the positive and negative provisions of our tax code is critical to structure your exploration deal properly.

This portion of your book will deal with the basic principles of Federal income tax law that affect the oil and gas industry. *Special structuring techniques* that can maximize your company's return through intelligent utilization of our tax laws, or can avoid unnecessary tax disasters, will be dealt with later in this book. See, e.g., § § **1.03(b)[8]**, **2.07**, and **3.04(b)[2]** *et seq.*

The special tax rules relating to oil and gas exploration and development primarily relate to the deductibility of *Intangible Drilling and Development Costs*, frequently referred to as "Intangibles" or "IDCs" (see § **1.03(b)[6]**), and the ability of the owner of an "economic interest" in oil and gas rights[18] to receive a portion of his revenues from the production of oil and gas tax free, due to the availability of a *Depletion Allowance* (see § **1.03(b)[4]**). In addition, where an oil and gas lease or oil and gas property has been held for more than 12 months, any gain on the sale of that lease or property will normally be taxed at Long Term Capital Gain rates (see § § **1.03(b)[2][d]** and **1.03(b)[7]**). However, this availability of Long Term Capital Gain treatment is restricted by the rules concerning the Recapture of certain tax benefits received from deductions for Productive-Well Intangible Drilling and Development Costs, prior Depreciation Deductions, and from Investment Tax Credits. (See § **1.03(b)[7]**.) Likewise, if the assignment of leases would be characterized as a sublease rather than as a sale, Long Term Capital Gain treatment will not be available, and the entire

18 "Economic interest" is defined as any interest in the minerals "in place" acquired by the taxpayer by investment, where the taxpayer must look to income from the extraction of the minerals for a return of his investment.

proceeds of the disposal (and not merely the "gain" element) will be subject to taxation at Ordinary Income rates. (See § **1.03(b)[7]**.)

Various of the tax consequences which apply to oil and gas transactions are determined on a *Property-by-Property* basis. The determination of what constitutes a "Property" involves some of the most complicated rules under the Federal Income Tax Code. Contiguous parcels received in a single conveyance or in simultaneous conveyances from the same owner constitute a single Property, although they still *may* be considered separate Properties as to separate mineral deposits.[19] However, parcels conveyed by separate owners are separate Properties, even though they involve contiguous (or even the same) tracts and were conveyed concurrently.

If a party owns dissimilar interests in a single tract, such interests constitute separate Properties. While all "operating interests" within a single tract will be aggregated unless the taxpayer files a timely election to treat the interests as separate Properties, non-operating interests are not aggregated. Furthermore, a subsequent transfer can convert what once was a single Property into separate Properties. (See § **3.04(b)[2][b]**.)[20]

A number of costs will be incurred by your company while leases are being held to determine whether or not the leases cover an area which appears to be prospective for oil and gas. Other costs will be incurred after an oil and gas property has been developed. General Administrative Overhead, salaries paid to your employees, costs of operating productive wells, and delay rentals paid following your acquisition of leases are all currently deductible as ordinary and necessary Business Expenses. However, from the standpoint of structuring your exploration deal, the primary influence of the Federal income tax laws will be the differing tax treatment of the following costs:

---

[19] In determining whether or not parcels are "contiguous," *cornering* is *not sufficient.*

[20] Different rules are applied in determining what constitutes a "Property" for purposes of the Crude Oil Windfall Profit Tax. See § **1.03(b)[4]**.

1. Lease Costs
2. Equipment Costs
3. Intangible Drilling and Development Costs

Lease Costs must be "capitalized" for Federal income tax purposes and become the taxpayer's *Depletable Basis* in the related leases. Such costs can be "recovered" for tax purposes in one of three ways:

1. As an *Abandonment Loss Deduction* if the lease is abandoned or otherwise condemned prior to establishing production.
2. Through the *Depletion Allowance* if the lease is productive.
3. As a *Reduction of Gain on Sale* if the lease or producing property is sold.

Equipment Costs must be "capitalized" for Federal income tax purposes, and become the taxpayer's *Depreciable Basis* in the equipment. These expenditures can likewise be recovered for tax purposes in one of three ways:

1. Through a *Depreciation Deduction* over the life of the equipment, or a statutorily-established *Recovery Period.*
2. Through an *Abandonment Loss Deduction* in the amount of the Adjusted Basis of the equipment at the time the equipment is abandoned.
3. As a *Reduction of Gain on Sale* at such time the equipment is sold, either separately, or as a part of the sale of the related oil and gas property.

Intangible Drilling and Development Costs, under a special option of the Code, may be *currently deducted* for Federal income tax purposes by any taxpayer owning operating rights in the related property.

§ 1.03(b)[4] Lease Costs, Delay Rentals, Depletion, and the Windfall Profit Tax. *Lease Costs* include not only bonuses paid for the acquisition of leases, but all costs related to lease acquisition, such as payments to lease brokers, recording fees, cost of title examination, etc. Lease Costs also include most Geological and Geophysical ("G & G") expenditures. In this regard, Bottom-Hole Contributions on productive wells are also treated as G & G expenditures, and thus as Lease Costs; and it is the position of the Internal Revenue Service that similar treatment should be extended to most Dry-Hole Contributions and Bottom-Hole Contributions on dry holes. (See § **2.07**.)

Lease Costs do *not* include Delay Rentals paid after the lease has been assigned to the paying party.

Expenditures which must be characterized as Lease Costs have the least value for Federal income tax purposes. For the major oil company, such expenditures at least increase the major's Depletable Basis in the leases and thus the amount of Cost Depletion it can subsequently claim. However, even for the major, this is far less valuable than the tax treatment that would arise if the expenditures were characterized as Equipment Costs, or, better still, as Intangible Drilling and Development Costs, where such costs could either be currently deducted or amortized on an "accelerated" basis over a fairly short period of time.

For a party who is entitled to claim *Percentage Depletion*—see below—the characterization of an expenditure as a Lease Cost will normally mean that such expenditure will have *no value* for Federal income tax purposes.

Obviously, expenditures which clearly relate to the acquisition of leases must be characterized as Lease Costs despite the unfortunate tax results. However, when purchasing a producing property, there is often a great degree of latitude in allocating the purchase price between the oil and gas reserves (which would be considered a Lease Cost) and the equipment (Equipment Costs). Of greater importance for the purposes of structuring your exploration deal, exploration agreements are often inadvertently structured in such a manner that expenditures which in fact relate to the acquisition of equipment

(Equipment Costs), or the drilling of wells (Intangible Drilling and Development Costs), must be recharacterized as Lease Costs. (See **§ 1.03(b)[6]**.)

Similarly, certain deal structures currently in use unnecessarily require Delay Rentals to be capitalized as Lease Costs rather than be currently deducted. (See **§ 3.04(b)[2][c-4-b]**.) While such capitalization may be of relatively minor significance in those areas where Delay Rentals are still a dollar an acre, the loss of such current deductions in Southern Louisiana and other areas where Delay Rental payments are quite high, often annually approximating the amount of the initial cash bonus, can have a serious impact on the economics of the deal.

As discussed below, while Lease Costs can be recovered, for tax purposes, through an Abandonment Loss deduction if a lease is surrendered or condemned prior to establishing production,[21] or through a reduction of gain on sale, the most common method of recouping Lease Costs on productive properties is through the *Depletion Allowance.*

Any revenues which can be received tax free by a taxpayer are far more valuable—actually *twice* as valuable—as revenues characterized as *taxable income.* Accordingly, the availability of the Depletion Allowance to the oil and gas industry has a very real bearing on the economics of any prospect.

Natural resources are "wasting assets," which, as they are produced, are being used up. Thus, every dollar received from the production of a natural resource represents, in part, a return of capital and not true "income." Since the taxpayer's investment in the property was not a deductible expenditure—see **§ 1.03(b)[2][c]**—this "return of capital" element of revenues received from production

---

[21] It is difficult, for tax purposes, to establish that a lease has been condemned by the drilling of a dry hole unless the lease is also surrendered. Likewise, if a taxpayer abandons a *productive well* prior to recovery of his entire Depletable Basis through Cost or Percentage Depletion, he cannot claim an Abandonment Loss, since the tax code requires that he claim the *greater* of Cost or Percentage Depletion, *properly calculated.* See below.

should not be subject to tax.

As to all natural resources, the owner of an "economic interest" in that asset is entitled to claim *Cost Depletion*—to amortize his Depletable Basis in the property over the life of the production.[22] However, to encourage producers to search for valuable minerals, the Internal Revenue Code also provides that the producers of certain minerals may compute an alternative *Percentage Depletion*, based on a statutorily-established percentage of the *gross income* from the property for that year.[23] Then, the producer, on a Property-by-Property basis, is entitled to claim (and *must* claim) the *greater* of Cost or Percentage Depletion for that year.

The Cost Depletion allowance is limited to the taxpayer's Depletable Basis in the property, and thus amounts to nothing more than a return of capital.[24] Furthermore, the calculation of Cost Depletion involves "spreading" the related Lease Costs over the recoverable reserves discovered to determine a "Per-Unit Cost"—the Lease Cost per recoverable barrel/MCF discovered—and then receiving back tax-free that portion of the sales price of that barrel of oil (or MCF of gas) which merely represents a recovery of those capitalized Lease Costs.[25] Cost Depletion, then by no stretch of the imagination constitutes a "tax "incentive"—it is merely a recognition of the economic "fact of life" that a mere return of capital is not true "income" or "gain," and shouldn't be taxed.[26]

---

[22] The "economic interest" concept is discussed in § **1.03(b)[3]**, n. 18.

[23] What constitutes a "Property" is discussed in § **1.03(b)[3]**.

[24] Compare the tax treatment of a sale or taxable exchange, where only the "gain" element is subject to taxation.

[25] The formula for calculating Cost Depletion is set forth in **Figure 2**.

[26] Note that the *price* at which the oil and/or gas is sold does not enter into the Cost Depletion calculation. Likewise, each year your remaining recoverable reserves ("Units") are recalculated, so that if your initial reserve calculation was inaccurate (and when isn't it?), you can (and must!) make adjustments in subsequent years.

Now, what about Percentage Depletion?

Until 1969, all members of the oil and gas industry were entitled to claim a Percentage Depletion allowance of 27½% of the *gross income* from the property. However, this Percentage Depletion rate for oil and gas was reduced to 22% of the *gross income* from the property by the Tax Reform Act of 1969.[27]

While the Tax Reform Act of 1969 significantly reduced the economic benefits of the Percentage Depletion allowance, it was left to the Tax Reduction Act of 1975 to put the final nails in the coffin, at least for the major oil companies. The 1975 Act repealed the availability of Percentage Depletion for the oil and gas industry with one important exception: a holder of an economic interest in oil and gas rights who qualifies as an "Independent Producer or Royalty Owner" can still claim Percentage Depletion, *so long as* the property was not "proven" at the time of its acquisition by the taxpayer.

It is obviously quite important to know who *is*—and who is *not*—an "Independent Producer" (or Royalty Owner).[28] Independent Producers and Royalty Owners, generally speaking, are owners of oil and gas rights who are not engaged in substantial refining or retailing activities.[29]

Even for the Independent Producer or Royalty Owner, the Percentage Depletion rate was reduced to 20% commencing in 1981, with further reductions to 18% in 1982, 16% in 1983, and 15% in 1984 and thereafter; and was limited to a specified annual Depletable

---

[27] "Excess" Percentage Depletion—the excess of Percentage Depletion over the taxpayer's Depletable Basis in the Property—also was made an "Item of Tax Preference" for both corporate and individual taxpayers by the Tax Reform Act of 1969, and thus could create Minimum Tax liability.

[28] As to the exact definition, see IRC, § 613A.

[29] Percentage Depletion is also based upon gross revenues *before deduction* of the Crude Oil Windfall Profit Tax. Thus, particularly as to "Tier One" and "Tier Two" production, Percentage Depletion could constitute a significant percentage of an Independent Producer's *net* (and otherwise taxable!) revenues.

Quantity of, as of 1980 and thereafter, 1,000 barrels of oil a day (or a 6,000 MCF/da. equivalent quantity of natural gas). However; since this allowance is based on the *gross income* from the property, it amounts to a significantly greater percentage of the *net* (i.e., taxable) income, after the deduction of operating costs.[29]

The availability of Percentage Depletion is of extreme importance to the Independent Producer. For instance, if Lease Costs were $64,000 on a prospect whose discovery well established 128,000 barrels of recoverable reserves to the Working Interest, the "Per-Unit Cost" for Cost Depletion Purposes would be 50¢ per barrel. If first year "Units Produced" were 25,600 barrels, $12,800 of working interest revenues could be received tax free under the Cost Depletion allowance. **(See Figure 2.)** However, if oil was selling at a "removal price" of $30.00 a barrel,[30] the Percentage Depletion allowance would be $115,200, assuming a 15% Percentage Depletion rate—or **nine times** than the tax-free revenues received by the producer who was limited to Cost Depletion![31] And note that, unlike Cost Depletion, the greater the sales price of the production, the greater the Percentage Depletion (and the tax-free share of revenues) will be.

Why are the tax-free revenues under Percentage Depletion normally so much greater than Cost Depletion? First, unlike Cost Depletion, Percentage Depletion is not limited to the taxpayer's Depletable Basis in the property. Thus, Percentage Depletion does provide a real incentive to the Independent Producer to engage in an active oil and gas search.[32] Second, the tax-free recovery permitted under Cost Depletion does not cover the drilling party's full cost of

---

[30] Percentage Depletion and the Windfall Profit Tax are both based on the "removal price" of the production—the price at which the production is (or would) have been sold "in the immediate vicinity of the well."

[31] In the initial year of drilling, the Independent Producer would, in all likelihood, still be limited to Cost Depletion due to the "50% of taxable income" limitation. See below.

[32] Note that the *Proven Property Exclusion* is aimed at making sure that this incentive motivates the independent producer to *drill* for oil and gas, and not merely to purchase producing properties.

UNRECOVERED DEPLETABLE BASIS

Divided by

REMAINING RECOVERABLE "UNITS"

Equals

PER-UNIT COST

Times

UNITS PRODUCED

Equals

COST DEPLETION ALLOWANCE

Figure 2: Cost Depletion Formula

drilling and equipping the well, but only his *Depletable Basis* in the property—his cost of *lease acquisition.*

As mentioned above, the amount of revenues which can be received tax-free has a significant impact on the economics of a prospect. If you are a drilling party who is entitled to claim Percentage Depletion, it is critical that the deal be structured in such a manner that Percentage Depletion is not inadvertently lost. Even if you are a major oil company which cannot claim Percentage Depletion, it is important that Percentage Depletion is not lost to any Independent Producer involved in the deal. (Since the availability of Percentage Depletion improves the economics for this Independent Producer, you can now structure a more favorable deal to *your company,* while providing the same economic return to the Independent, so long as he can still claim Percentage Depletion. See § **1.03(b)[1]**. Unfortunately, under many current deal structures, Percentage Depletion is unnecessarily being lost for Independent Producers participating in the drilling. (See § **3.04(b)[2][c-4-b]**.)

There are certain restrictions on the availability of Percentage Depletion, even to the Independent Producer or Royalty Owner. The *Depletable Quantity* limitation has already been discussed. We have also mentioned the *Proven Property Exclusion.*[33] Furthermore, Percentage Depletion is not available as to foreign production, and is limited to 50% of the *taxable income* from the property. Since the year's IDC deductions are taken into account in calculation this limitation, Percentage Depletion normally *will* exceed the 50% of taxable income limit in the year the initial well on a property is drilled. Thus, for that year, the drilling party normally must claim Cost Depletion, even if he is an Independent Producer.[34]

---

[33] The definition of "proven property" for purposes of the Proven Property Exclusion does not follow general industry rules and is quite complicated and cloudy. See Houghton (ed.), *Miller's Oil and Gas Income taxation,* ¶ 31-3-4.

[34] Independent Producers and Royalty Owners are subject to a further limitation that Percentage Depletion cannot exceed 65% of the *taxpayer's taxable income* for the year.

Any consideration of the tax treatment of production revenues, and any evaluation of the economics of a given prospect, must also take into account the effects of the Crude Oil Windfall Profit Tax of 1980. This grossly-mislabeled tax obviously operates to the detriment not only of the oil and gas industry, but of the general pubic well. However, its existence is an economic fact of oil and gas exploration and development. The tax will hereinafter be referred to as the "Windfall Profit Tax" or "WPT."

The so-called "Windfall Profit" element on which the WPT is levied is the difference between the *removal price* (generally speaking, the "selling price")[35] of domestically-produced crude oil, and an *adjusted base price*, separately set for three different "Tiers" (types) of production. Tax rates also vary depending on the Tier of oil produced; and Independent Producers are taxed at reduced rates on Tier 1 and Tier 2 (but not Tier 3) *working interest* production up to an aggregate Tier 1/Tier 2 production of 1,000 barrels a day.[36]

While the tax is limited to domestically-produced "crude oil," *crude oil* is defined to include condensate and distillate, so that for all practical purposes, only dry gas production is excluded from the impact of the Act.

Obviously, the definition of "Tier" is a critical factor in determining what impact the WPT will have on the economics of a proposed deal. The Tiers are generally defined as follows:

1. *Tier 1 Oil* is *oil discovered prior* to 1979, other than oil from stripper wells.

---

[35] See note 30, above, and Burke and Bowhay, *Income Taxation of Natural Resources*. ¶¶ 31.18-31.19.

[36] The Act contains severe restrictions to prevent transfers from non-qualified to qualified Independent Producers, or restructuring from non-working interests to working interests to play games with the tax rate.

The Crude Oil Windfall Profit Tax Act is a highly-complicated piece of legislation, and the discussion in this section will be purposely oversimplified. For detailed discussion, see Burke and Bowhay, *Income Taxation of Natural Resources*, Chapter 31; and Houghton (ed.), *Miller's Oil and Gas Federal Income Taxation*, Chapter 32.

2. *Tier 2 Oil* is *oil produced from stripper well properties.*[37]

3. *Tier 3 Oil* is *newly-discovered oil,* i.e., oil produced from a Property from which no oil was produced in calendar year 1978.[38]

The base price for Tier 1 Oil is $12.81, adjusted quarterly for inflation since June, 1979. (Remember, the lower the base price, the more revenues subjected to Windfall Profit Tax.) The tax rate on the Windfall Profit from such Tier 1 Oil is 70% for major oil companies and Royalty Owners; however, Independent Producers are taxed at a reduced rate of 50% on the Windfall Profit element of revenues from their working interest production (subject to the aggregate Tier 1/Tier 2 limitation mentioned above).

Tier 2 Oil is assigned a base price of $15.20 a barrel, adjusted for inflation and certain quality and location differences. The general tax rate is 60%, with Independent Producers being entitled to claim a reduced rate of 30% on their working interest production. (However, see below as to the impact of ERTA on stripper well production of Independent Producers.)

The base price for Tier 3 Oil is set at $16.55, adjusted for inflation *plus two percentage points,* and, again, for quality and location differences. The tax rate for all producers, including Independent Producers, is 30%.

Since, in evaluating the economics of a prospect the "Tier" classification of any crude oil to be produced is a critical factor, it is equally critical to realize that the definition of "Property" for purposes of the WPT does not follow the normal "Property" concept

---

[37] Also included in Tier 2 Oil is *oil produced from a National Petroleum Reserve* in which the United States has an economic interest. Under ERTA, "exempt stripper well oil" of an Independent Producer is exempt from the WPT. See below.

[38] Tier 3 Oil also includes *heavy oil* (oil with an API specific gravity of 16 degrees or less) and *incremental tertiary oil.* Under ERTA, that portion of Tier 3 oil which qualifies as "newly-discovered oil," will be taxed at further reduced rates, commencing in 1982. See below.

discussed in § **1.03(b)[3]**, but follows the rules promulgated by the Department of Energy for crude oil price control purposes. Under the DOE definition, a lease constitutes a single "Property," *despite a subsequent assignment of a separated portion of that lease.* While each separate and distinct producing reservoir may be treated as a separate "Property," Farmouts and other assignments of rights under a "single right to produce" will not necessarily create new "Properties" for WPT purposes if the wells are to be drilled to the same, currently-producing horizon.[39]

Obviously, in structuring an exploration deal, economics will also vary, based on whether or not a given participant in the deal will be taxed as an "Independent Producer." The definition of "Independent Producer" follows the same definition utilized under the Independent Producers and Royalty Owners Exemption discussed for Percentage Depletion purposes under the Tax Reduction Act of 1975. (See above.) No similar reduced rate exists for "Royalty Owners."[40] However, as a result of ERTA, "Qualified Royalty Owners" are exempt from the Windfall Profit Tax on two barrels of *Qualified Royalty Production* a day, commencing in 1982, with the exemption increasing to three barrels a day for 1985 and thereafter.[41] Furthermore, for 1983 and thereafter, stripper oil production received by an Independent Producer is exempt from the Windfall Profit Tax, so long as the property was not owned, at any time after

---

[39] The normal "Property" definition under the IRC does apply to the computation of the *net income* limitation. See below.

[40] Certain credits for "Qualified Royalty Owners" were, however, subsequently created.

[41] A "Qualified Royalty Owner" is a producer who is an individual, an estate, or a qualified family farm corporation. *Qualified Royalty Production* means crude oil attributable to an economic interest other than operating interest. See IRC, § 6429(d).

As in the case of the rules concerning the Independent Producers and Royalty Owners Exemption under the Tax Reduction Act of 1975, this exemption will be lost upon the transfer of an interest in a "proven property." Likewise, related royalty owners are only entitled to one exemption, which is allocated among them in proportion to their share of total production.

July 22, 1982, by any person other than an Independent Producer; and, commencing in 1982, the Windfall Profit Tax rate for all parties is significantly reduced on Tier 3 Oil, which qualifies as "newly-discovered oil" though not on other classes of Tier 3 oil, with an eventual leveling out at a 15% rate for 1986 and thereafter.

A number of other special rules apply in computing the WPT. The tax is limited to 90% of a producer's *net income* from the property, and may further be reduced, under certain circumstances, by certain state severance taxes on the "Windfall." The inflation adjustment is based on the GNP Deflator; and, as an excise tax, the WPT is a *deductible business expense* for Federal income tax purposes. Likewise, as mentioned above, the Windfall Profit Tax does not reduce *gross income* from the property for purposes of computing Percentage Depletion.[42]

While an eventual phaseout of the WPT is provided for in the Act, many producers are skeptical as to whether this phase out will ever occur.

§ 1.03(b)[5] Equipment Costs, ACRS and Other Depreciation Deductions, and the Investment Tax Credit. Just as Lease Costs must be capitalized and recovered through deductions for Cost or Percentage Depletion, the cost of *depreciable personal property* must be capitalized, and becomes the taxpayer's *Depreciable Basis* in the property to be recovered primarily through *Depreciation Deductions.*[43]

While costs of such salvagable tangible property as tubing, recoverable casing, drilling rigs, separators, and tank batteries must, like Lease Costs, be capitalized rather than expensed, such

---

[42] Remember, the greater the *gross income* from the property, the greater the amount of tax-free revenues under the Percentage Depletion Allowance.

[43] As to the possible recovery of Equipment Costs through Abandonment Loss Deductions or as a reduction of gain on sale, see **§ 1.03(b)[3]**.

Equipment Cost expenditures are much more valuable, for tax purposes, than expenditures for Lease Costs. First, Depreciation Deductions can be claimed *in addition* to Percentage Depletion—a significant benefit to the Independent Producer. (See § **1.03(b)[4]**.) Furthermore, for Independent Producers and major oil companies alike, the ability to claim Depreciation Deductions on an "Accelerated" basis, and, in addition, to take advantage of the Investment Tax Credits often created by such expenditures, create definite tax advantages. Furthermore, ERTA has in many instances improved the tax treatment available for Equipment Costs by providing liberalized Depreciation and Investment Tax Credit rules.

As to Depreciation, ERTA replaces prior recovery methods with a new method called the *Accelerated Cost Recovery System* ("ACRS"). Under this system, *Recovery Property* (depreciable property placed in service after 1980) is classified as three-year, five-year, ten-year, or 15-year Recovery Property. (Most of the Equipment Costs with which you will be dealing will be characterized as three-year or five-year Recovery Property.) The cost of such property may then be recovered over the specified period, using statutory tables which provide a method of Accelerated Amortization built into the statute.[44]

Under the ACRS method, the *entire cost* of depreciable property can be recovered over the statutory period, as contrasted with the prior Depreciation method which required that the amount recovered through Depreciation be reduced by the estimated salvage

---

[44] As to Accelerated versus Straight-Line (ratable) Amortization, see § **1.03(b)[2][c]**. The accelerated method includes a so-called "half-year convention." However, taxpayers who so desire can utilize the straight-line (ratable) method of Depreciation rather than the statutory accelerated method.

ERTA also permits a taxpayer, as a replacement for the prior *Bonus Depreciation* allowance, to annually expense a limited dollar amount of equipment expenditures.

Unlike the prior system, under ERTA the same accelerated methods of cost recovery are available, and an identical recovery period is established, whether the property is new or used.

value of the property.[45]

The Internal Revenue Code also creates certain "tax credits" to encourage expenditures which the Government feels is socially desirable. (See § **1.03(b)[2]**.) One such incentive is an "Investment Tax Credit" ("ITC") of a specified percentage of the cost of depreciable tangible personal property acquired by a taxpayer for use in his trade or business. This tax credit is normally available in connection with expenditures for Equipment Costs.

ERTA has significantly changed and liberalized the rules concerning the availability of the Investment Tax Credit. Under ERTA, eligible five-year Recovery Property qualifies for an ITC equal to 10% of the cost of such property.[46] (As a *tax credit,* this 10% allowance is a *direct offset* against taxes payable, and thus twice as valuable as an equivalent amount of *tax deduction.*) Furthermore, prior rules concerning the limited amount of ITC on *used personal property* in any taxable year has been liberalized.[47]

The full Investment Tax Credit is allowed in the year property is "put in service," without any proration for the period of time during the year that the property is in service. However, the ITC is limited to property predominantly used in the United States, or used in developing natural resources from ocean waters.

Section **1.03(b)[2]** discussed the potential abuses perceived by Congress in permitting a taxpayer to take rapid write-offs and Investment Tax Credit benefits against Ordinary Income upon the acquisition of property, which subsequently could be sold at Long Term Capital Gain rates. As mentioned, in many instances, to

---

[45] Prior rules which could mean that such accelerated recovery methods would result in a potential Add-On Minimum Tax Liability for taxpayer's who were not "Regular Business Corporations" have been significantly changed by TEFRA.

[46] A 6% tax credit applies to three-year Recovery Property.

[47] The dollar limit on the cost of such used property is raised to $125.00 for 1981 through 1984, and $150,000 thereafter.

prevent such *Long Term Capital Gains Conversion*, Congress provided for any gain on the sale of such property to be "Recaptured" at Ordinary Income (rather than Long Term Capital Gains) rates to the extent of all or part of prior Depreciation Deductions and Investment Tax Credits upon the sale of the property.

The Congressional "Recapture" policy carries over to the area of Equipment Costs. As under prior law, ACRS deductions are fully recaptured upon the sale of Recovery Property, to the extent of any gain on the sale.[48] ERTA also provides for a recapture of prior ITC benefits if the property is disposed of prior to the end of the three - or five-year Recovery Period. However, unlike the prior law, these recapture rules have been substantially liberalized to reduce the amount subject to recapture by two full percentage points of credit for each full year the property is held before disposition. Thus, if five-year Recovery Property is disposed of within a year, the full 10% Investment Tax Credit is subject to recapture. However, if the Property is not disposed of until after one full year, the amount of the credit subject to recapture is reduced to 8%, to 6% after two years, to 4% after three years, and 2% after four years. There is no recapture if the property is held for five full years.

§ 1.03(b)[6] Intangible Drilling and Development Costs. While ERTA has significantly improved the tax treatment of Equipment Costs, expenditures for Intangible Drilling and Development Costs continue to receive the most favorable treatment under our Federal income tax laws.

Intangible Drilling and *Development Costs* ("Intangibles" or "IDCs") are defined under the Internal Revenue Code as costs which

---

[48] The recapture extends not only to the prior recovery deductions, but also to any portion of the cost of the property which was expensed under the ERTA option rather than depreciated. See note 44, above.

Note that, unlike the rule discussed in note 45 of this subsection concerning Items of Tax Preference, the recapture is not limited to the "excess" of Accelerated over Straight-Line Depreciation.

in themselves have no salvage value, and which are necessary and incident to the drilling of wells or the preparation of wells for production.[49]

(Generally speaking, if you can't kick it, and it doesn't relate to leases, it's an Intangible!)

Normally, expenditures included within the definition of Intangible Drilling and Development Costs would be required to be capitalized as a permanent betterment of the land. However, under a special election provided under the Internal Revenue Code, a taxpayer who owns operating rights in an oil and gas property may elect, at his option, "expense," rather than capitalize, his expenditures for Intangibles.

Figure 3 illustrates the significance of the taxpayer's right to expense, rather than capitalize, Intangibles. As shown, the deductibility of the Intangibles reduces the after-tax cost of the well for the 50% bracket taxpayer to *half* of what the cost would have been if the IDCs had been capitalized.

"TEFRA"—the Tax Equity and Fiscal Responsibility Act of 1982—affected the current deductibility of Intangible Drilling and Development Costs for corporations which do not qualify as Independent Producers. Such "Integrated Oil Companies" are now limited, under § 291(b) of the Code, to a current deduction for only 85% of their IDCs, but are permitted to amortize the remaining 15% ratably over a 36-month period, beginning in the month in which the costs are paid or incurred.

While the new TEFRA IDC rules will cause substantial IDC deductions to be deferred by "Integrated Oil Companies," the limited amount of IDCs required to be capitalized, and the rapid

---

[49] Note that, while in connection with Lease Costs, expenditures *related to the acquisition* of leases are themselves treated as Lease Costs, services rendered in connection with the installation of equipment, etc., are characterized as IDC expenditures, and not Equipment Costs, so long as the services are necessary and incident to the drilling of the well or the preparation of the well for production.

	IDC Expensed	IDC Capitalized
1. Well Costs	$100,000	$100,000
2. Tax Savings	[50,000]	-0-
3. **After-Tax Cost of Well**	**$ 50,000**	**$100,000**

*Assumes 50% Federal income tax bracket

Figure 3: Economic Impact of Deductibility of Intangibles

amortization, are far less serious economically than might first appear. Accordingly, for majors and Independents alike, IDC deductibility remains a key to well economics.

The IDC deduction cannot be disproportionate to the operating rights which will be owned by the taxpayer at the time he achieves "complete payout" of his costs of drilling, completing, equipping, and operating the well. Deductibility will not be affected by the ownership by another party (such as a Farmor) of a *non-operating interest* (e.g., an overriding royalty interest or net profits interest) during the Payout period, even if such interest is convertible to an operating (working) interest after Payout, or by a reserved "reversionary" working interest which is subordinated to "complete payout" to the drilling party. However, if the drilling party must "carry" another party for an interest in the well, or bears, at any stage during the drilling, etc., a share of well cost disproportionate to his pre-payout working interest, or if another party owns a reversionary interest which could revert prior to "complete payout," the portion of drilling costs disproportionate to the drilling party's *assured* pre-payout working interest *must be capitalized as a Lease Cost.*[50]

As illustrated in **Figure 4**, the impact of a partial loss of IDC deductibility on the economics of a prospect is very serious. As **Figure 4** shows, the cost of such a well is 50% higher for the drilling party if only one-half of the IDCs are deductible than if the full IDC deduction were available. Furthermore, the characterization of such capitalized drilling costs as *Lease Costs*—the least favorable type of exploration expenditure for tax purposes—makes the capitalization even more costly. Accordingly, most exploration deals are structured

[50] The IRS has even taken the position that if there is *any possibility* that the taxpayer's share of well costs may be disproportionate to his share of pre-payout revenues, or that an interest could revert prior to complete payout, the potentially disproportionate share of well costs must be capitalized as a Lease Cost, irrespective of whether or not a disproportionate pre-payout sharing actually occurs.

Differing types of interests which a Farmor might reserve are discussed in § **3.04(b)[3]** *et seq.*

	IDC 100% Deductible	IDC 50% Deductible
1. Well Costs	$100,000	$100,000
2. Tax Savings	[25,000]	50,000
3. **After-Tax Cost of Well**	**$ 75,000**	**$ 50,000**

*Assumes 50% Federal income tax bracket

Figure 4: Economic Impact of Partial Loss of IDC Deductibility

to permit the drilling party to take a deduction for the full amount of his Intangible Costs. (See § **3.04(a)**.)

As discussed in § **3.04(b)[2][c-4-b]**, prior structuring techniques, which appeared to "lock" the parties to a Farmout into essentially one method of structuring, have since proved an unnecessary strait-jacket. There are methods of structuring deals in which the parties can "have their cake and eat it too" through more economically-desirable structuring methods without a loss of IDC deductibility for the drilling party. Furthermore, certain structuring methods which have been used despite a perceived loss of deductibility for the drilling party, such as the celebrated "third for a quarter" arrangement, can now be utilized with minor variations in structuring without any deductibility loss. (See § **3.04(b)[2][c-5-b-4]**.)

IDC deductions can be taken pursuant to *"Turnkey" arrangements*, where a fixed price is paid to the drilling contractor for the drilling of the well, even though the actual costs incurred by the drilling contractor are significantly less, so long as certain structuring guidelines are taken into account.[51] Likewise, under certain circumstances, drilling costs can be *prepaid* by a cash basis taxpayer for a well which will not be drilled until the following year; if the prepayment is properly structured, the taxpayer will still be permitted to take the deduction in the year of payment, rather than in the year in which the well is actually drilled.[52] However, where such arrangements are entered into with the promoter of a deal, or where the promoter or an affiliate serves as the "drilling contractor," the IRS has taken a position that such arrangements will be viewed with suspicion.[53]

---

[51] See Burke and Bowhay, *Income Taxation of Natural Resources*, ¶ 14.03; Houghton (ed.), *Miller's Oil and Gas Federal Income Taxation*, ¶ ¶ 20-21, 20-26.

[52] See Burke and Bowhay, *Income Taxation of Natural Resources*, ¶ 14.11; Houghton (ed.), *Miller's Oil and Gas Federal Income Taxation*, ¶ 20-10.

[53] See Rev. Rul. 80-71.

The disposal of an oil and gas property on which producing wells have been drilled may result in a potential recapture of certain of the prior tax advantages created by IDC deductibility—see § **1.03 (b)[7]**.

§ **1.03(b)[7] Disposing of Oil and Gas Properties.** A final consideration in analyzing how Federal income tax consequences affect the economics of your deal structure is to understand what happens, tax-wise, when an oil and gas property is disposed of.

Whether your company is disposing of an undeveloped lease or a producing property the assignment will, for tax purposes be treated as falling into one of four categories:[54]

(1) Sale
(2) Exchange[55]
(3) Sublease
(4) Sharing Arrangement.

The tax consequences of your trade may vary markedly, depending on the category into which the transaction falls.

When *property is traded solely for cash*, the transaction is treated, for tax purposes, as a *Sale*. The selling party is taxed on the "gain" element only as to the compensation he receives—the difference between his *amount realized on sale* (the cash paid) and his *Adjusted Basis* in the property he has disposed of. See § **1.03(b)[2][d]**.

If the property has been held for over 12 months by the selling party, and is not an *Ordinary* (i.e., inventory) *Asset*, the gain will be taxed, subject to the recapture rules discussed below, at the more favorable Long Term Capital Gains rates. Otherwise, the higher Ordinary Income tax rates will apply. See § **1.03(b)[2][d]**.

---

[54] Whether the property is producing or undeveloped will, however, affect whether or not the recapture consequences discussed below will come into play.

[55] Exchanges may be further subdivided into *taxable* and *non-taxable* Exchanges. See below.

For reasons previously discussed—see § **1.03(b)[2][d]**—Congress is not too happy with the games that can be played by taking deductions in connection with a property at Ordinary Income tax rates in one year, and then disposing of the property a year later, when any resulting "gain" on the sale would be taxed only as a Long Term Capital Gain. To restrict perceived abuses, the Tax Code often provides that, upon the disposal of a property, resulting gain may be taxed ("Recaptured") at Ordinary Income tax rates, and certain other previously-claimed tax benefits must be restored, to the extent certain tax benefits specified in the Code were previously claimed.

Where developed oil and gas properties are involved, any gain on sale of depreciable equipment will be recaptured at Ordinary Income tax rates to the extent of all prior ACRS and other Depreciation deductions. Prior Investment Tax Credits may also be subject to recapture, in whole or in part, when the equipment has been held for less than five years. See § **1.03(b)[5]**.

An even more serious recapture potential exists when a producing property is disposed of. A portion of the IDC deductions previously claimed in connection with the property are now also subject to recapture.[56] (If dry holes have been drilled on the same property on which the producing wells are located, these "dry hole IDCs" must also be included in the recapture calculation.)[57]

Where *property is traded for property*, the transaction, for tax purposes, is considered an *Exchange*.[58] Generally, most Exchanges

[56] The "Recapture" calculation is quite complicated, and confusion exists concerning just how much of the prior "Productive Well IDCs" are subject to recapture for an Independent Producer who has claimed Percentage Depletion in connection with production from the Property. See Burke and Bowhay, *Income Taxation of Natural Resources*, ¶ ¶ 14.20 *et seq.*; Houghton (ed.), *Miller's Oil and Gas Federal Income Taxation*, ¶ 20-37.

[57] As to what constitutes a "Property," see § **1.03(b)[3]**.

[58] As to "mixed" transactions—a trade of property for property *plus cash*, see Burke and Bowhay, *Income Taxation of Natural Resources*, ¶ 5.03.

are treated as "taxable events"; there will be a tax payable on the "gain" element: the difference between the *amount realized on Exchange*—the Fair Market Value of the property as of the date of the Exchange—by each exchanging party and his Adjusted Basis in the property. Whether Ordinary Income or Long Term Capital Gains rates apply, and the application of any recapture rules, will depend on the same guidelines which applied to Sales. Each exchanging party also "steps up" his Basis in the property received to the Fair Market Value as of the date of the Exchange on which he was taxed. See § **1.03(b)[2][d]**.

Certain Exchanges of property can, however, be made *tax-free*. These non-taxable Exchanges include exchanges of business or investment property for business or investment property of a like kind (e.g., a lease for a lease); the exchange by a corporation of its stock for cash or property; and the exchange with a partnership of cash or property for an interest in that partnership. However, upon making such a non-taxable Exchange, the exchanging parties are not permitted to "step up" their Basis, but must "carry over" their existing Basis in the property given up, which now becomes their Basis in the newly-acquired property. See § **1.03(b)[2][d]**.

But what if you *trade an oil and gas property for cash* ***plus a reserved interest in production***, such as an Overriding Royalty Interest or a Net Profits Interest? The transaction is taxed as a *Sublease* rather than as a Sale, with several unfortunate consequences to the transferor.[59] First, the *entire amount* realized on the Sublease, and not merely the "gain" element, is subject to tax. (Your Basis isn't used to reduce the amount subject to tax, since it remains your Basis in the reserved interest.) Second, no matter what the nature of the asset, and no matter how long you have held it, the tax is always levied at Ordinary Income tax rates. (Whether an Independent

[59] Considerable controversy exists over whether or not retention of certain types of *Carried Working Interests*, or the carving out of Override prior to the transfer (or the assignment back, rather than the reservation, of the interest in production) would avoid *Sublease* characterization.

Producer can claim Percentage Depletion on the amount received is a presently raging controversy.)[60]

But what if you *trade property for the transferee's agreement* to *assume a part of the burden of developing that property?* The transaction is a *"Sharing Arrangement,"* which creates no present taxable consequences to either party to the deal.

In case you didn't catch it, the "Sharing Arrangement" just mentioned is "legalese" for the bread-and-butter Farmout Agreement, where the Farmor assigns an interest in the Farmout Leases to the Farmee in return for the Farmee's commitment to develop the Farmout Leases. This classification of the Farmout as a *"Sharing Arrangement"* for tax purposes is one reason for its frequent utilization as an exploration tool—neither party's economics is adversely affected by any current taxes payable on the deal.

Unfortunately, the favorable classification of the Farmout as a Sharing Arrangement requires that the Farmee's commitment relate to the same property that is to be assigned him by the Farmor. Otherwise, the transaction becomes an *"Exchange of Property for Services"* for tax purposes—with highly unfortunate tax consequences to both parties to the deal!

(More of this when we discuss the nightmare of *Revenue Ruling 77-176* in § **3.04(b)[2]** *et seq.*)

§ 1.03(b)[8] Impact of Tax Laws on Deal Structure-General Principles. As indicated in § **1.03(b)[1]**, the tax rules discussed above simply touch the surface of the complicated impact of our Federal income tax laws on structuring exploration deals. Specific structuring considerations, and a more detailed discussion of certain tax rules, are included later in this book. However, some general

---

[60] All producers will be permitted an allowance for Cost Depletion on the otherwise taxable amount. Where "rank wildcat" acreage is involved, this could have the same effect as subjecting only the "gain" element to tax, although the tax would still be levied at Ordinary Income tax rates. See *Collum v. U.S.*, 45 A.F.T.R. 80-751 (1979).

observations concerning the impact of our Federal income tax laws on deal structure can and should be made here.

As emphasized in § **1.03(b)[1]**, the tax consequences of an exploration deal are just as much an element of return or cost as the price at which the production is sold or the charges made by the drilling contractor. The **tax positives** of the arrangement will **improve** return; the **tax negatives** will **reduce** it.

In selecting among various deal alternatives, then, the varying tax implications of the various structures must be taken into account. Obviously, there are benefits in selecting a deal structure which takes advantage of tax positives, and avoids (or at least reduces) tax negatives. Certain of these "tax enhancements" can be secured at no "cost" to you; these clearly are to be taken advantage of. On the other hand, certain deal structures which have attractive tax consequences carry a heavy "price tag" in the form of non-tax disadvantages. (See, e.g., § **3.04(b)[2][c-5-b-4]** and its discussion of the disadvantages of the Tax Partnership approach.)

As was also discussed in § **1.03(b)[1]**, tax enhancement is not an automatic goal in itself; it is simply **one factor** to be taken into account in analyzing the advantages and disadvantages of available structuring alternatives. Where the structure which maximizes tax considerations has non-tax negatives, the relative advantages and disadvantages must be *quantified* to arrive at an intelligent conclusion as to whether or not the tax advantages outweigh the non-tax "price tag."[61]

Analyzing the *value* of the tax consequences of various structuring techniques requires a consideration of the following:

1. The **amount** involved—e.g., *the larger a deduction or tax credit, and the smaller the amount of taxable income or gain subject to recapture, the better.*

---

[61] For an example of such quantifying, see § **3.04(b)[2][d]**.

2. The **timing** of the tax benefit or detriment. In this regard, consider:

a. **When** will the benefit or detriment occur? *The quicker the better for tax benefits*, such as deductions or tax credits—e.g., *accelerate "taxable" losses; the slower the better for tax detriments*—in other words, *try to defer taxable income, gain, or recapture.*

b. On *amortized or "installment" items:*

(1) **What is the "Recovery (amortization) Period?"** *The shorter the better for taxable losses, etc.; the longer the better for items of taxable income or gain.*

(2) **How rapidly *within* that period** will the loss be taken or the income or gain reported? *Accelerated Amortization gives greater value to taxable losses* than does a Straight-Line Amortization. (The opposite is true for taxable income or gain.)

3. The **rate** at which taxable income or gain will be taxed, or at which otherwise taxable income which is being offset by taxable losses *would* have been taxed. In other words:

a. *The lower the rate at which income or gain will be taxed, the better* (try to realize taxable revenues as a Long Term Capital Gain, rather than as Ordinary Income);

b. *The higher the rate at which income offset by deductions would otherwise have been taxed, the more valuable the deduction* (e.g., a deduction should be used to offset Ordinary Income whenever possible, rather than merely reducing or eliminating a Long Term Capital Gain).

Let's end our discussion of the impact of tax laws on deal structure by posing three structuring "axioms":

1. ***Never structure a deal so that a party will unnecessarily lose a deduction.*** As discussed above, in certain circumstances you may select a structure which is less favorable, tax-wise, because of its more advantageous economic impact. However, to lose deductions ***unnecessarily*** for either you or the other party to the deal—without an overriding gain in economic benefit—makes no sense whatsoever.

2. ***Take into account the Time Value of Money and its impact on tax laws.*** This can be done by taking deductions as early as possible, and deferring the recognition of income as long as possible.
Time Value of Money principles are discussed in § **1.03(b)[2][c]**. While a few limited opportunities exist for permanently avoiding a tax,[62] most tax structuring principles merely involve the deferring of a tax liability.

3. ***Avoid, whenever possible, the payment of taxes on "Phantom Income."***[63] To pay a tax before money is actually received unnecessarily reduces cash flow and earnings per share for the major oil company, and normally proves disastrous for the cash-poor and borrowed-up Independent. Furthermore, such payment of taxes on "paper profits" often results in taxes on an "assumed" economic gain that will never be realized, and may even amount to characterizing a transaction as profitable for tax purposes where, in economic reality the parties suffered an economic loss.

---

[62] Examples are Percentage Depletion and Long Term Capital Gains Conversion. See § § **1.03(b)[2][d]** and **1.03(b)[4]**.

[63] See, e.g., § **3.04(b)[2][a]**.

§ 1.03(c) Principles of Petroleum Economics*

§ 1.03(c)[1] In General. It is easy to lose track of the *actual goal* of your company: what leads it to engage in the search for oil and gas. *Finding oil* is NOT your company's objective; its objective is to *make money*. The *method* by which this goal is accomplished is to search for oil and gas on a profitable basis.

Your company may have various subsidiary objectives. If it is a major oil company whose stock is listed on the New York Stock Exchange, stability of income, availability of cash for the payment of dividends, increased earnings per share under generally accepted financial accounting principles, and a growth in the value of the stock will all be important considerations. To the Independent, an increase in the reserves in the ground, not reflected in taxable income, may be a more important goal. Similarly, sufficient presently-available cash flow to satisfy the Independent's constant loan repayment requirements and to make distributions to investors is critical to the survival of the company. However, for oil companies, operation *at a profit* must be the primary objective.

A company cannot survive unless it is operating profitably. This means that, in addition to adding to its reserves in the ground, the company must generate sufficient available cash to meet its general operating expenses, to satisfy its loan repayment (debt service) obligations, to meet shareholder/investor goals as to dividends or cash distributions, and to have sufficient funds left over to undertake new projects.

Finding oil is not particularly difficult: there are many areas in the United States where wells can be drilled which seldom or ever result in a dry hole. However, finding oil *profitably* is another matter entirely.

For an Independent who raises his drilling capital from outside

* The author's thanks go to Patrick W. Fitzgerald for his assistance in the preparation of this portion of the book.

investors, the announcement of various productive wells may keep those investors happy for a period of time, even though their overall investment or even the majority of their productive wells are not returning reasonable, or any, profits. Similarly, the exploration department of a major oil company may be able to excite management with announcements of various major "finds," even though overall operations—when dry holes, marginal wells, and cost overruns are rolled in—are not providing an adequate return. However, at some point in time, the realization that operations are not being conducted profitably will sink in for the parties who control the purse strings.

In the absence of blind and unlikely luck, to achieve the goal of profitable operation the explorationist must establish the minimum Rate of Return necessary to satisfy the profitability requirements listed above, and must then select projects which, based upon a reasonable and realistic economic analysis, appear likely to attain this minimum return objective. A failure to analyze potential profitability recklessly endangers your company's financial survival—and, if your management is intelligent, your own job longevity!

§ 1.03(c)[2] Measuring Project Profitability

§ 1.03(c)[2][a] General Principles. The importance of measuring the potential profitability of a project has been mentioned in § **1.03(c)[1]**. If the projects available to your company do not appear likely to generate sufficient profits to attain your minimum return objectives as outlined in that section, an undertaking of such projects makes no economic sense; your company would be better off to invest its money in money market instruments and wait out the storm. However, most members of the oil and gas industry have more projects available to them which should attain or exceed the necessary minimum return objectives than their available capital will permit them to handle. Under this more common circumstance, measuring potential project profitability helps your company *maximize return* by making an intelligent selection among competing opportunities.

Far too many people, both inside and outside the oil and gas industry, measure profitability based on the gross revenues to be received, without taking into account the *length of time* necessary for the receipt of those revenues. However, each time you spend a dollar, you are losing the *Time Value* of the money — your ability to invest it elsewhere and receive a return on that investment. Likewise, the sooner you receive funds, the more valuable they are to you, since you can now put those funds to work and receive a return on your investment. Thus, in analyzing the estimated cost of a project and the anticipated revenues, both expenditures and revenues of the project must be *discounted* to allow for this Time Value of Money. Similarly, the *timing/rapidity* of return may be just as (or even more) important than the gross revenues to be received from the project, or the price at which the product will be sold.[64]

To properly take into account the Time Value of Money — to calculate, as an element of expense, the loss of use of the money invested in the project until revenues are received, and the overall discounting effect to be applied to future expenditures and receipts of revenue — an appropriate *discount factor* must be selected. Selecting a discount factor which is too low will make projects appear much more potentially profitable than they are in economic reality; on the other hand, imposing too high a discount factor will make it similarly difficult to justify undertaking any drilling opportunity.

In selecting the appropriate discount factor, you, or your company's petroleum economist, will take into account current and anticipated interest and inflation rates, as well as the estimated return you could realize on your capital if you invested it in other

[64] As to the importance of Time Value principles as applied to the impact of tax laws on the structuring of exploration deals, see § **1.03(b)[8]**.

projects available to you. For instance, if you know you could make a 10% to 15% return on investment virtually risk free merely by depositing your company's funds in "Money Market" certificates, undertaking projects which will not "break even" at this discount rate makes no sense, unless you anticipate a future reduction in the rate of return on "Money Market" certificates, or suspect that your revenues from the project on which your project economic analysis has been based have been significantly understated.

The higher the appropriate discount rate, the more important the Time Value principle becomes. In the 1950s, when the inflation rate stood at 2%, and the prime interest rate at 4%, the failure to take Time Value into account in estimating project economics was of relatively little significance. However, to ignore Time Value today, in the face of current inflation and interest rates, is nothing short of irresponsible.

Sample discount factors are shown at **Figure 5.**[65]

Any analysis of project economics must be based on *after tax calculations*. The significance of the incentives—and negatives—which the Internal Revenue Code provides for oil and gas exploration and development has been discussed in § **1.03(b)[1]**. Since only those dollars which your company is permitted to keep after giving Uncle Sam his share are available for satisfying company obligations, and undertaking additional operations (plus paying or increasing your salary!), analysis based on *pre-tax* figures can be highly misleading.

Full-blown analysis of project economics by a petroleum economist will include calculations of Time Value and tax impact, and may thus be quite complicated. However, when analyzing project economics, you can perform "Quick and Dirty" analyses by using rules of thumb which have, at least to some extent,

---

[65] These discount rates assume that the monies are received or spent at the *end of the applicable year*.

NUMBER OF YEARS	1%	2%	3%	4%	5%	6%	7%	8%	9%	10%	12%	14%	15%	16%	18%	20%
1	.9901	.9804	.9709	.9615	.9524	.9434	.9346	.9259	.9174	.9091	.8929	.8772	.8696	.8621	.8475	.8333
2	.9803	.9612	.9426	.9246	.9070	.8900	.8734	.8573	.8417	.8264	.7972	.7695	.7561	.7432	.7182	.6944
3	.9706	.9423	.9151	.8890	.8638	.8396	.8163	.7938	.7722	.7513	.7118	.6750	.6575	.6407	.6086	.5787
4	.9610	.9238	.8885	.8548	.8227	.7921	.7629	.7350	.7084	.6830	.6355	.5921	.5718	.5523	.5158	.4823
5	.9515	.9057	.8626	.8219	.7835	.7473	.7130	.6806	.6499	.6209	.5674	.5194	.4972	.4761	.4371	.4019
6	.9420	.8880	.8375	.7903	.7462	.7050	.6663	.6302	.5963	.5645	.5066	.4556	.4323	.4104	.3704	.3349
7	.9327	.8706	.8131	.7599	.7107	.6651	.6227	.5835	.5470	.5132	.4523	.3996	.3759	.3538	.3139	.2791
8	.9235	.8535	.7894	.7307	.6768	.6274	.5820	.5403	.5019	.4665	.4039	.3506	.3269	.3050	.2660	.2326
9	.9143	.8368	.7664	.7026	.6446	.5919	.5439	.5002	.4604	.4241	.3606	.3075	.2843	.2630	.2255	.1938
10	.9053	.8203	.7441	.6756	.6139	.5584	.5083	.4632	.4224	.3855	.3220	.2697	.2472	.2267	.1911	.1615
11	.8963	.8043	.7224	.6496	.5847	.5268	.4751	.4289	.3875	.3505	.2875	.2366	.2149	.1954	.1619	.1346
12	.8874	.7885	.7014	.6246	.5568	.4970	.4440	.3971	.3555	.3186	.2567	.2076	.1869	.1685	.1372	.1122
13	.8787	.7730	.6810	.6006	.5303	.4688	.4150	.3677	.3262	.2897	.2292	.1821	.1625	.1452	.1163	.0935
14	.8700	.7579	.6611	.5775	.5051	.4423	.3878	.3405	.2992	.2633	.2046	.1597	.1413	.1252	.0985	.0779
15	.8613	.7430	.6419	.5553	.4810	.4173	.3624	.3152	.2745	.2394	.1827	.1401	.1229	.1079	.0835	.0649
16	.8528	.7284	.6232	.5339	.4581	.3936	.3387	.2919	.2519	.2176	.1631	.1229	.1069	.0930	.0708	.0541
17	.8444	.7142	.6050	.5134	.4363	.3714	.3166	.2703	.2311	.1978	.1456	.1078	.0929	.0802	.0600	.0451
18	.8360	.7002	.5874	.4936	.4155	.3503	.2959	.2502	.2120	.1799	.1300	.0946	.0808	.0691	.0508	.0376
19	.8277	.6864	.5703	.4746	.3957	.3305	.2765	.2317	.1945	.1635	.1161	.0829	.0703	.0596	.0431	.0313
20	.8195	.6730	.5537	.4564	.3769	.3118	.2584	.2145	.1784	.1486	.1037	.0728	.0611	.0514	.0365	.0261
25	.7798	.6095	.4776	.3751	.2953	.2330	.1842	.1460	.1160	.0923	.0588	.0378	.0304	.0245	.0160	.0105
30	.7419	.5521	.4120	.3083	.2314	.1741	.1314	.0994	.0754	.0573	.0334	.0196	.0151	.0116	.0070	.0042
40	.6717	.4529	.3066	.2083	.1420	.0972	.0668	.0460	.0318	.0221	.0107	.0053	.0037	.0026	.0013	.0007
50	.6080	.3715	.2281	.1407	.0872	.0543	.0339	.0213	.0134	.0085	.0035	.0014	.0009	.0006	.0003	.0001
60	.5504	.3048	.1697	.0951	.0535	.0303	.0173	.0099	.0057	.0033	.0011	.0004	.0002	.0001	*	*

DISCOUNT RATE

*The factor is zero to four decimal places.

Figure 5: Discount Factor Table

taken tax and Time Value implications into account.[66] Thus, a "*Return on Investment*" ("ROI") measure of profitability can (and should) be criticized for ignoring Time Value effect (see § **1.03(c)[2][b]**), and is often calculated by the non-economist on pre-tax figures. However, by doing a quick calculation of *normal* Time Value and tax impacts, you can come up with a "quick and dirty" ROI standard which, more often tha not, will tell you whether you want to give a proposed project further consideration.

Obviously, the profitability of a project must be determined over its whole life, and not upon any current "slice of time." This principle is referred to as "Full Cycle Economics." Thus, to analyze the profitability of a drilling venture in the year in which the well was drilled would show that the project was operating at a loss, unless the well was capable of "paying out" in 12 months or less. On the other hand, to analyze the well during the following year, thus ignoring the cost of drilling, might indicate a profitable venture when in fact the well would never return its initial cost of drilling, even on an after-tax basis.

Despite the principle of Full Cycle Economics, before undertaking a project you must also make sure that *current* operations will generate sufficient available cash to meet your company's current cash flow needs, such as satisfying operation expenses, debt service, etc. (See § **1.03(c)[1]**.) No matter how profitable certain projects may appear, to strap yourself for available cash denies you the flexibility to take advantage of even more profitable opportunities that may arise in the future, and, when carried to an extreme, can endanger your

[66] One such handy test in the Time Value area is the "Rule of 72." A clever economist noted that, at all discount rates, dividing the assumed Time Value/discount rate into 72 would give the approximate number of years which it takes your money to double (or to be cut in half by inflation), with almost the same accuracy that a formal discounting provides. Thus, at a 10% Time Value/discount rate, money doubles (or its buying power is halved) in 7.2 years; at 15%, the doubling/halving occurs in 4.8 years; and at 20%, only 3.6 years is required. (At 2%, it took 36 years!)

company's financial survival through current insolvency.[67] Similarly, binding commitments which drastically restrict your company's future flexibility of operation may prevent you from taking advantage of next year's profitable opportunities. Thus, before funds are permanently tied up for years to come, or arrangements such as fixed price contracts, area calls on production, or far-reaching areas of mutual interest agreements are entered into, you must balance the value to your company of this present arrangement with the loss of "Opportunity Value" in the future.

What if you have undertaken a project based on certain assumptions that later prove untrue? (And who periodically doesn't?) Two-thirds of your estimated expenditures have been made; it is obvious that the project will never be more profitable. However, without some additional expenditure, no revenues will be received from the project whatsoever.

An analysis of whether or not to continue such a project depends on an evaluation of " 'Yet to Spend' Economics." The "Sunk Costs"are ignored; and project economics are valued on the dollars yet to be spent versus the anticipated return if *those dollars* are expended. If, based upon this "Yet to Spend" analysis, continuation of the project meets your return objectives, the project should be continued. On the other hand, if even on a " Yet to Spend" basis the project offers an inadequate return, it should be junked rather than continuing to throw good money after bad.[68]

---

67 But can't monies always be borrowed against future reserves? While such *Production Loans* are of critical importance to the continued operation of most Independents, a total reliance on the future availability of borrowing fails to take into account that, in an uncertain economy, Production Loans may be unavailable irrespective of the value of the reserves in the ground. Also, in an inflationary period, interest rates may climb so high that the economics of the project are now destroyed.

68 In analyzing a "Yet to Spend" proposal, it should be remembered that the person who originally proposed the project normally feels under great pressure to justify its continuation. Accordingly, it is critical to carefully analyze the realism with which the dollars yet to be spent, and the estimated return to be realized if those dollars are spent, has been projected.

The foregoing discussion assumes that we know exactly how many dollars a project will cost, the volume of reserves which will be discovered, and what portion of those reserves will be recoverable, the price which the production will bring, the time over which this money will be received, the number of productive wells, etc. In the real world, however, each of these factors is uncertain and must be *estimated.* Furthermore, since petroleum economics are always analyzed on an after-tax basis (see above), we are forced to assume that existing tax laws will continue unchanged. This assumption, while necessary, has clearly been proved by past experience to be fallacious. Accordingly, our estimates of project profitability must also involve adjustments to take into account the *degree of probability* that each of our assumptions are accurate ("Risk"), plus what to do with those assumptions where *no degree of probability can be estimated* ("Uncertainty").

§ 1.03(c)[2][b] Measures of Profitability. Various formulas are used in attempting to measure the anticipated profitability of a proposed petroleum venture. Some of these approaches are easy to apply but give limited assistance in analyzing project profitability, serving primarily as "rule of thumb" measurements. (The "answer" provided by such analysis can be highly misleading.) Others are extremely sophisticated; while they are more difficult to calculate, the answers provided by such approaches are far more useful.[69]

Just as you may have difficulty in communicating with the tax specialist due to the complexity of the tax laws and his specialized jargon, it is often difficult for ordinary human beings to effectively communicate with the petroleum economist. Certain of the necessary approaches used in measuring project profitability are quite

[69] Don't be lulled into thinking that any of these approaches provide guaranteed accuracy as to the profitability of a *particular project.* Of necessity, such analysis will be based on numerous assumptions, few of which will be 100% accurate. See § **1.03(c)[3]**, dealing with adjustments for Risk and Uncertainty.

complicated; and the petroleum economist also has a language all his own. Likewise, economists (and company managements!) have been known to "fall in love with the numbers," forgetting that experience and common sense show that "economic evaluations" only represent an informed guess as to *possible* project results.

Despite this, it is critical that you understand the basic principles that the petroleum economist applies in analyzing a prospect, so that you can communicate with him and understand what he is saying—and can make your own preliminary determinations concerning which projects (and deal structures) should be given serious consideration by your company. While no one expects you to be able to apply the more complicated measure of profitability approaches, a general understanding of what these measurements are all about will permit you to do "Quick and Dirty" studies to determine whether or not a given prospect should be considered by your company at all, and if so, which method of structuring the deal should prove the greatest benefit to your company. (See **§ 1.03(c)[2][a]**.)

For purposes of illustrating various measures of profitability, let's assume you are considering four possible projects:

1. *Project A* is a proposed development well which should result in steady production. Following an initial $10,000 expenditure, it is expected that you would receive $40,000 in net revenues, realized evenly over a ten-year period.

2. *Project B* is a development well with an anticipated early flush production, followed by a rapid decline, leveling out to long-lived production. The well will cost $10,000; you expect to realize $7,500 in Year One, $2,500 in Year Two, and $1,000 a year for the next 60 years, for an aggregate return of $70,000.

3. *Project C* is an exploratory well in a remote area. The well, again, will cost $10,000, but is expected to return $50,000 (adjusted for risk). However, production is not expected to commence until Year Six; and the $50,000 net revenues are expected to be realized evenly over a five-year period in Years

Six through Ten.

4. *Project D* is a project involving a high abandonment expenditure. It will require an initial $5,000 expenditure in Year One, followed by an additional $5,000 expenditure at the end of Year Ten. The aggregate $40,000 return will be realized 25% ($10,000) in Year One, 50% ($20,000) in Year Two, and the balance ($10,000) equally over the next eight years.

In many businesses, the common profitability standard is "Mathematical Return." This is a simple mathematical calculation arrived at by *dividing the expected return by the number of years over which the return will be realized*, then expressing this as a *percentage of the investment* to arrive at an *annual return figure.*

The simplest example of the Mathematical Return calculation is to look at a $100 savings account, which earns $5 in interest over a 12-month period. The Mathematical Return is [5% ($5 ÷ 1) ÷ $100 = 5%].

As illustrated in **Figure 6**, based on a Mathematical Return analysis, Project C is our best bet, and Project B our worst.

The obvious flaw in such an evaluation is that it does not take into account *when* the monies are to be received. Thus, a project which returns $910 on a $1,000 investment in Year One, with an additional $10 a year over the next ten years; and a project which, on a similar investment, returns $1,000 in Year Ten, would each have a Mathematical Return of 10% a year. (Compare, in **Figure 6**, Projects A and D, where there is a significant difference in the timing of the expenditures and revenues, but an identical Mathematical Return.)

Fortunately, Mathematical Return is not a common method of evaluating oil economics.

Two methods of measuring profitability which are common in the oil industry are *"Return on Investment"* ("ROI") and *"Payout"* ("PO").

The petroleum economist defines *"Return on Investment"* as the *cumulative net cash flow over project life, divided by the maximum cash outlay.* Simplified, ROI means *what* you anticipate *you are going to get out of the project, divided by what you'll have to put in.*

FORMULA:

(Expected Return ÷ No. of Years Over Which Return Realized) ÷ Investment

PROJECT A

($40,000 ÷ 10) ÷ $10,000 = **40%**

PROJECT B

($70,000 ÷ 62) ÷ $10,000 = **11.29%**

PROJECT C

($50,000 ÷ 10) ÷ $10,000 = **50%**

PROJECT D

($40,000 ÷ 10) ÷ $10,000 = **40%**

Figure 6: Mathematical Return Example

As illustrated in **Figure 7**, under the ROI method, Project B (the worst project for Mathematical Return) is the best investment, Project C is the next best, and Projects A and D rank identically (and last).

ROI is the darling of the promoter talking to prospective investors. It is also widely used since it is relatively easy to calculate. However, ROI shows *magnitude* of return only, without giving any consideration to *timing*. Thus, a "4 to 1 Deal" might or might not be acceptable to your company, depending on whether the 4 to 1 return was to be received over a two-year or a 20-year period, and further depending on how much of the return was to be received early in project life (thus, the high ranking of Projects B and C, and the "identical" rankings of Projects A and D.)

"Payout" is technically defined by the petroleum economist as the *length of time for the cumulative cash position of the project to reach zero.* Simply stated, it's *how quick you'll get your money back.* Again, Payout is relatively easy to calculate and thus is a popular measurement approach when describing a project to a layman. **Figure 8** shows the Payout ranking with Project D (six months) coming in first, Projects B (two years) and A (2.5 years) second and third, and Project C (six years) last — which had ranked first under Mathematical Return and high under ROI — however, Payout considers *time only*, without any consideration of the *magnitude of the return* — what will be received after payout — or the *timing* of that *after-payout return. (Thus, if Project D had gone to water in the seventh month, it would still rank better on a PO basis than Project A.)*

ROI and PO analyses do have their uses. First, since they lend themselves to a fairly rapid calculation, and involve substantially less assumptions than more calculated analyses, they are a good, quick test

---

[70] Of course, the usefulness of the analyses also depends upon the realism of the assumptions. See **§ 1.03(c)[3]**.

FORMULA:

Cash Flow Divided by Cash Outlay
equals Return on Investment

PROJECT A $40,000 ÷ $10,000 = **4:1 ROI**

PROJECT B $70,000 ÷ $10,000 = **7:1 ROI**

PROJECT C $50,000 ÷ $10,000 = **5:1 ROI**

PROJECT D $40,000 ÷ $10,000 = **4:1 ROI**

Figure 7: Return on Investment (ROI) Example

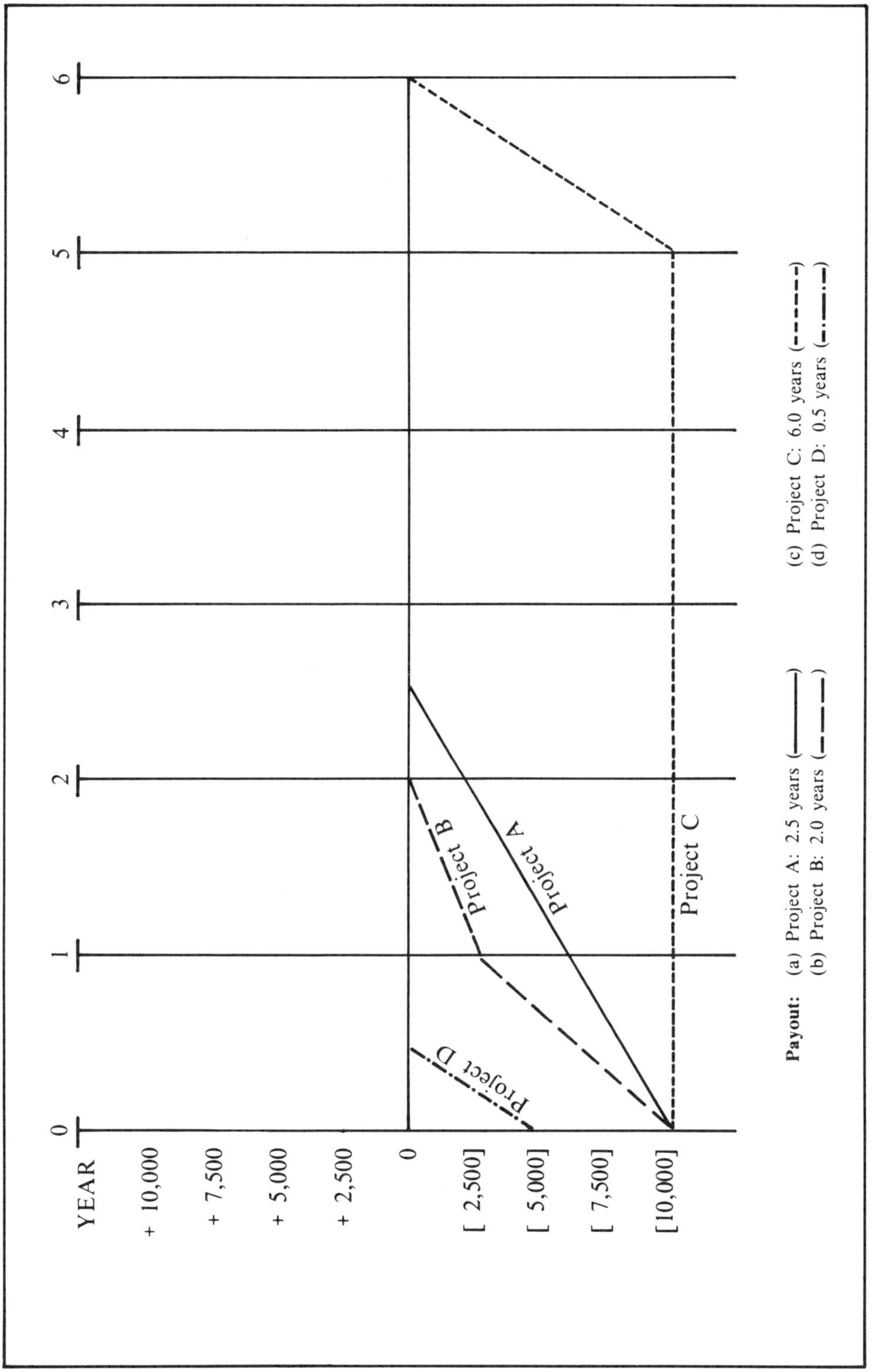

Figure 8: Payout (PO) Example

for *rejecting* a proposed undertaking.[71] (A project that would attain Payout in eight years and return 1.5 to 1 on your company's investment is unlikely to prove very exciting to management.) Likewise, in making a preliminary analysis, an ROI calculation coupled with certain basic assumptions concerning the timing of the return, e.g., initial cash flow will commence within nine months, and the total return will be realized fairly evenly over ten years, can be of some use. Similarly, a quick Payout means that your initial investment will have been recovered at least within a limited period of time if the project is successful and can thus be used to undertake additional ventures. Conversely, an extended payout period means that your investment will face much greater exposure to changing political and economic factors that are difficult to evaluate, so that there will be more uncertainty concerning the accuracy of the assumptions used in making your projections.[72]

Now let's look at certain methods of analysis that are apt to give a more accurate profitability measurement. The most commonly-applied "sophisticated" analysis method is *"Internal Rate of Return"* ("IRR"). *"Internal Rate of Return"* is technically defined as the *discount rate which will make the present worth of the project's net cash flow equal zero.* Simply stated, it's an *"effective" Rate of Return* — an annual Rate of Return *discounted for time.*

The computation of IRR is considerably more complicated than the ROI or PO calculations, but has a far greater degree of usefulness. An estimation is made of the revenues that will be received and the expenditures made for each year of the project's life. The discount rate, which, when if applied to each of these figures would result in a "zero" present worth, is then calculated. This

---

71 See § **1.03(c)[2][a]**.

72 It may prove relatively simple to determine whether or not major oil companies might be nationalized within the next two years. But what about the next ten? Likewise, an analysis of anticipated oil prices 12 months from now is more likely to prove relatively accurate than trying to project oil prices 25 years in the future.

calculation requires a trial and error approach when the analysis is being done manually; it can be performed almost instantaneously, however, through the use of a computer or sophisticated calculator.

The IRR analysis is the method chiefly favored by most major oil company managements, and certainly presents a much more accurate method of evaluation than ROI and PO. Thus, in IRR ranking, Project A emerges as by far the best of the projects. (See **Figure 9**.)

IRR has its own built-in fallacies, which, under certain circumstances, can lead to an unrealistic answer. First, it assumes that *cash* will be *reinvested at the same rate as the calculated IRR.* This means that the IRR approach makes a low return project appear even less favorable, and overstates the return from a highly-profitable project. Likewise, since many projects will move from negative to positive positions several times over the project life, the IRR analysis can provide *multiple answers* — several discount rates which would result in a zero present worth. (Project D is an example of this—the additional investment at the end of Year 10 creates two annual "Negative" Periods.) Furthermore, IRR does not take into account *magnitude of return.*

Another analysis method which provides useful information by taking into account Time Value considerations is a *"Present Worth"* ("PW") calculation, also referred to as *"Present Value"* or *"PV."*

As defined by the petroleum economist, Present Worth (Present Value) means *the sum of all future cash effects discounted at a stated rate.* For our purposes, we're talking about the *value, today, of amounts to be expended and received in the future.*

The PW calculation takes into account *both magnitude* and *timing* to arrive at a *dollar measurement* of anticipated project return. Obviously, the selection of the appropriate discount or "return" rate is of critical importance to the usefulness of the PW analysis. (See **§ 1.03(c)[2][a]**.) However, it is hoped your company will have a pretty good idea of the proposed return which it expects to achieve from its projects. By applying this as a "discount factor" both to

PROJECT A	IRR = **38.45%**
PROJECT B	IRR = **25.76%**
PROJECT C	IRR = **22.93%**
PROJECT D	IRR = **?**(1)

(1) No IRR can be computed for Project D: *see text.*

Figure 9: Internal Rate of Return (IRR) Example

future cash expenditures and future receipts, you will be able to realistically determine whether or not a given project should attain the desired return objective.[73]

Obviously, the value of a PW analysis depends not only on the accuracy of the necessary assumptions (see § **1.03(c)[3]**), but also on whether or not the selected "discount" or "return" factor is a realistic one. (See § **1.03(c)[2][a]**.) If the desired return is too high, few projects will achieve that return unless the project assumptions are equally unrealistic. On the other hand, an unrealistically low discount rate will lead to the acceptance of projects which will not in fact prove sufficiently profitable for your company, and a misranking of projects due to an inadequate consideration of the impact of Time Value.

Figures 10-12 illustrate the effectiveness of the PW analysis, and the impact of the discount rate selected. If a *2% discount rate* is used (**Figure 10**), Time Value counts for little (and every project will appear "profitable"): Projects B and C rank best; Project A and D rank approximately the same, and last.

With a *10% discount rate* — a 20% *pre-tax* return, since we're assuming our figures are presented on an *after-tax* basis (see § **1.03(c)[2][a]**) — all projects still make money: Project D is clearly the best project; Projects A and C rank approximately the same (with C now trailing); and Project B — our best "ROI" project — is *clearly* last. (**Figure 11**.)

At a *20% discount rate*, Project D (**Figure 12**) has totally outstripped the field (the value of its earlier returns and deferred investment coming to the fore); Project A is far ahead of Project C —

---

[73] If after making the PW calculation the sum of the discounted positive figures (the revenues) and the discounted negative figures (the expenditures) give a non-negative result (either positive or "breakeven"), the project should achieve, or exceed the desired return. If the sum of the totals is negative, the desired return objective presumptively would not be achieved by this particular project.

		PROJECT A		PROJECT B		PROJECT C		PROJECT D	
YEAR	DISCOUNT FACTOR	CASH FLOW	PW*	CASH FLOW	PW*	CASH FLOW	PW*	CASH FLOW	PW*
0	-0-	[10,000]	[10,000]	[10,000]	[10,000]	[10,000]	[10,000]	[5,000]	[5,000]
1	.9804	4,000	3,922	7,500	7,353	-0-	-0-	10,000	9,804
2	.9612	4,000	3,845	2,500	2,403	-0-	-0-	20,000	19,224
3	.9423	4,000	3,769	1,000	942	-0-	-0-	1,250	1,178
4	.9238	4,000	3,695	1,000	924	-0-	-0-	1,250	1,155
5	.9057	4,000	3,623	1,000	906	-0-	-0-	1,250	1,132
6	.8880	4,000	3,552	1,000	888	10,000	8,880	1,250	1,110
7	.8706	4,000	3,482	1,000	871	10,000	8,706	1,250	1,088
8	.8535	4,000	3,414	1,000	854	10,000	8,535	1,250	1,067
9	.8368	4,000	3,347	1,000	837	10,000	8,368	1,250	1,046
10	.8203	4,000	3,281	1,000	820	10,000	8,203	[3,750]	[3,076]
11-62	**	-0-	-0-	52,000	26,369	-0-	-0-	-0-	-0-
NET PRESENT WORTH			**$25,930**		**$33,167**		**$32,692**		**$28,728**

*Discounted at 2%

**Discount Factors for Years 11 through 62 at a 2% Discount Rate vary from .8044 in Year 11 to .2929 in year 62

Figure 10: Present Worth/Present Value (PW/PV) Example (at 2% Discount Rate)

		PROJECT A		PROJECT B		PROJECT C		PROJECT D	
YEAR	DISCOUNT FACTOR	CASH FLOW	PW*	CASH FLOW	PW*	CASH FLOW	PW*	CASH FLOW	PW*
0	-0-	[10,000]	[10,000]	[10,000]	[10,000]	[10,000]	[10,000]	[5,000]	[5,000]
1	.9091	4,000	3,636	7,500	6,818	-0-	-0-	10,000	9,091
2	.8264	4,000	3,306	2,500	2,066	-0-	-0-	20,000	16,528
3	.7513	4,000	3,005	1,000	751	-0-	-0-	1,250	939
4	.6830	4,000	2,732	1,000	683	-0-	-0-	1,250	854
5	.6209	4,000	2,484	1,000	621	-0-	-0-	1,250	776
6	.5645	4,000	2,258	1,000	565	10,000	5,645	1,250	706
7	.5132	4,000	2,053	1,000	513	10,000	5,132	1,250	642
8	.4665	4,000	1,866	1,000	467	10,000	4,665	1,250	583
9	.4241	4,000	1,696	1,000	424	10,000	4,241	1,250	530
10	.3855	4,000	1,542	1,000	386	10,000	3,855	[3,750]	[1,446]
11-62	**	-0-	-0-	52,000	3,827	-0-	-0-	-0-	-0-
NET PRESENT WORTH			**$14,578**		**$7,121**		**$13,538**		**$24,203**

*Discounted at 10%

**Discount Factors for Years 11 through 62 at a 10% Discount Rate vary from .3505 in Year 11 to .0027 in Year 62

Figure 11: Present Worth/Present Value (PW/PV) Example (at 10% Discount Rate)

		PROJECT A		PROJECT B		PROJECT C		PROJECT D	
YEAR	DISCOUNT FACTOR	CASH FLOW	PW*	CASH FLOW	PW*	CASH FLOW	PW*	CASH FLOW	PW*
0	-0-	[10,000]	[10,000]	[10,000]	[10,000]	[10,000]	[10,000]	[5,000]	[5,000]
1	.8333	4,000	3,333	7,500	6,250	-0-	-0-	10,000	8,333
2	.6944	4,000	2,778	2,500	1,736	-0-	-0-	20,000	13,888
3	.5787	4,000	2,315	1,000	579	-0-	-0-	1,250	723
4	.4823	4,000	1,929	1,000	482	-0-	-0-	1,250	603
5	.4019	4,000	1,608	1,000	402	-0-	-0-	1,250	502
6	.3349	4,000	1,340	1,000	335	10,000	3,349	1,250	419
7	.2791	4,000	1,116	1,000	279	10,000	2,791	1,250	349
8	.2326	4,000	930	1,000	233	10,000	2,326	1,250	291
9	.1938	4,000	775	1,000	194	10,000	1,938	1,250	242
10	.1615	4,000	646	1,000	162	10,000	1,615	[3,750]	[605]
11-62	**	-0-	-0-	52,000	806	-0-	-0-	-0-	
NET PRESENT WORTH			**$6,770**		**$1,458**		**$2,019**		**$19,745**

*Discounted at 20%

**Discount Factors for Years 11 through 62 at a 20% Discount Rate vary from .1346 in Year 11 to .00001 in Year 62

Figure 12: Present Worth/Present Value (PW/PV) Example (at 20% Discount Rate)

the prime "Mathematical Return" contender, and Project B.[74]

Let's discuss a final measure of profitability — *"Discounted Return on Investment"* ("DROI"). As in the ROI calculation, a DROI analysis involves *dividing how much you hope to receive from the project by how much you'll have to spend to get there;* unlike the ROI calculation, however, *both expenditures and revenues are discounted* to take into account Time Value.

Just as in the case of the PW analysis, DROI takes into account both *magnitude* and *time.* However, where the PW analysis gives us a profitability measurement in terms of a *dollar figure,* the DROI calculation — as illustrated in **Figure 13** — gives us a *ratio* of return to investment. Here, again, at a low discount rate, Projects B, C, and D rank virtually the same, with Project A trailing. (Project D ranks closer to B and C, even at a low discount rate, than it did in the PW calculation, due to its lower discounted investment.) However, at higher discount rates, Project D clearly becomes the best project (becoming even more impressive the higher the discount rate selected); Project A equals and then exceeds Project C; and Project B now trails.

See **Figure 14** for a graphic illustration of the impact of Time Value and discount rates on the ranking of these projects.

(Note that Project D's DROI remains relatively constant despite increases in discount rate, since revenues are received relatively early in project life, and a major part of the expected expenditures have been deferred to the future. This means that Project D will be less sensitive to discount rates, uncertainty, and unexpected change.)

There are many other methods by which the profitability of a proposed drilling venture can be analyzed. Any of these methods will be useful to you in determining which projects to undertake and

---

[74] Project B's only saving grace over Project C is its two-year Payout. This would have to be balanced against its lower Present Worth.

But *Quaere:* Is management being too optimistic in going to a 20% discount rate — a **40% pre-tax annual discounted return?**

FORMULA:

Discounted Revenue Stream divided by Discounted
Cash Outlay equals
Discounted Return on Investment

PROJECT A

(a) At 2%:	$35,930	÷ $10,000	=	3.6:1 DROI
(b) At 10%:	$24,578	÷ $10,000	=	2.5:1 DROI
(c) At 20%:	$16,770	÷ $10,000	=	1.7:1 DROI

PROJECT B

(a) At 2%:	$43,166	÷ $10,000	=	4.3:1 DROI
(b) At 10%:	$17,121	÷ $10,000	=	1.7:1 DROI
(c) At 20%:	$11,458	÷ $10,000	=	1.1:1 DROI

PROJECT C

(a) At 2%:	$42,692	÷ $10,000	=	4.3:1 DROI
(b) At 10%:	$23,538	÷ $10,000	=	2.4:1 DROI
(c) At 20%:	$12,019	÷ $10,000	=	1.2:1 DROI

PROJECT D

(a) At 2%:	$37,829	÷ $9,101	=	4.2:1 DROI
(b) At 10%:	$31,130	÷ $6,927	=	4.5:1 DROI
(c) At 20%:	$25,552	÷ $5,807	=	4.4:1 DROI

Figure 13: Discounted Return on Investment (DROI) Example

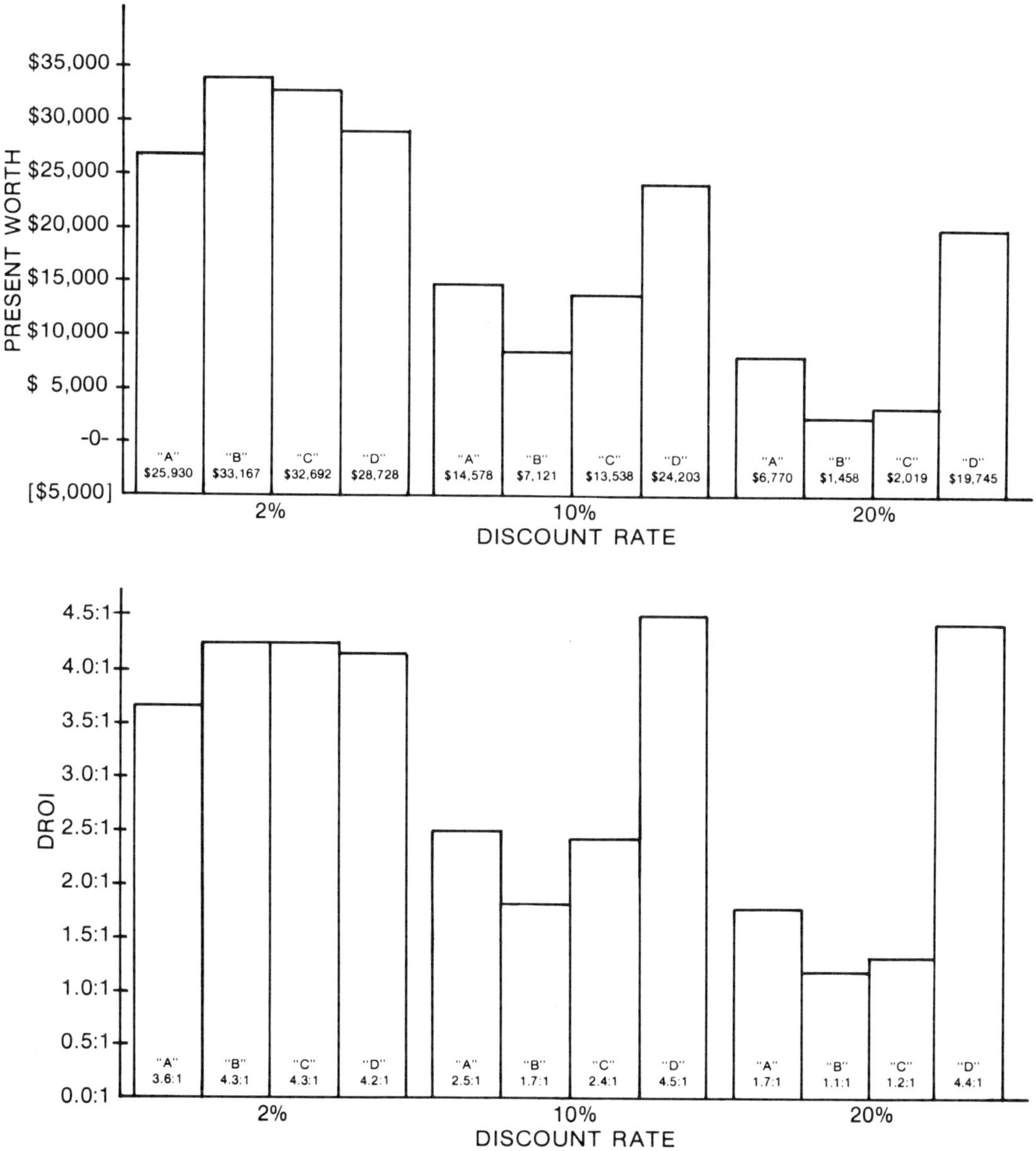

Figure 14: Graphical Illustration of Impact of Discount Rate on Time Value

which to reject, so long as they are *uniformly applied* and do not involve inherent fallacies, such as unrealistic assumptions or ignoring Time Value considerations. The critical point is that *some realistic method* of project evaluation be undertaken before determining, first, whether or not to undertake a project at all, and, second, which method of structuring the deal will prove most beneficial to your company.

Most successful explorationists have, in fact, been analyzing anticipated project profitability, even if only on a subconscious basis. (Those who don't usually go out of business fairly quickly when their string of luck runs out.) However, even experienced explorationists frequently neglect to analyze whether the profits from drilling "heads up" might not be bettered by some other type of arrangement, or whether profits under the deal structure proposed couldn't be improved by a more innovative structuring.

Many explorationists and company managements fall in love with a given prospect and become hooked on keeping 100% of the prospect for the company. However, it is inherent in exploratory drilling that a large number of prospects, no matter how high their quality, will be nonproductive or disappointing. Measuring the expected results of alternate methods of structuring such deals permits a businesslike decision as to whether "own account" drilling, support, farming out, or farming in should prove more advantageous from your company's standpoint. Such analyses make the oil and gas search an acceptance of *calculated* risks, rather than wild speculation.

§ 1.03(c)[3] Evaluating Risk and Uncertainty. The measures of profitability discussed in **§ 1.03(c)[2][b]** all assume that certainty exists as to the numerous factors which will affect the profitability of a given venture. However, as discussed in **§ 1.03(c)[2][a]**, this "assumption" is obviously untrue.

The oil and gas search involves numerous engineering and geological uncertainties. Will hydrocarbons be present in *any* quantity? What will be the thickness of the sand and the aerial extent of the structure? Will the hydrocarbons be recoverable, and if so, to what extent? What will rates of production be?

In addition to these technical uncertainties, there is also considerable uncertainty concerning the impact of political and economic conditions and changes on the feasibility of a project. What will the inflation rate be? What will be the impact of inflation, recession, and other economic factors on costs and the price at which any oil and gas produced can be sold? Will the political climate be pro-oil, anti-oil, or relatively neutral? What changes will occur in our tax laws?

All of these uncertainties must be taken into account in realistically evaluating the desirability of a given undertaking. This means that you must estimate "Risk" as to those factors where degrees of probability can be reasonably estimated,[75] and must further take into account "Uncertainty" as to those factors where a degree of probability of occurrence cannot be estimated.[76]

Obviously, it is hard to work an "Uncertainty Calculation" into an economic equation, since the degree of probability that the given event will or will not occur cannot be estimated. However, such "Uncertainty Exposure" must be taken into account in determining whether or not to undertake a project.

In conducting an Uncertainty Evaluation, economists frequently look at " 'Worst Case' Economics," i.e., what is the *most you can lose* if the "Worst Case Scenario" occurs?[77] Another approach is to evaluate

---

[75] E.g., whether or not a given development well will result in a dry hole.

Note that "Risk" refers to the number of different possible outcomes and their likelihood of occurrence, not to their desirability or non-desirability. Thus, jumping out of an airplane without a parachute at 10,000 feet is "risk free," because there is only one possible outcome. Jumping out of the same airplane when it is located on the ground is a "risky possibility" — you could land unhurt, twist your ankle, or break your leg.

[76] E.g., whether or not petroleum will be replaced by an alternate cheap and equally-desirable energy source, whether or not the industry will be nationalized; the likelihood of a near-term war in the Mid-East.

[77] This is referred to as calculating your *Maximum Negative Cash Position.*

the cost of taking the next step, such as shooting seismic, renewing leases or paying delay rentals, rather than permitting the leases to expire.[78]

1982 brought home to the industry just how "Risky" and "Uncertain" economic projections can be. Projects undertaken in the late 1970s and early 1980s had been justified on the then apparently-unassailable assumptions that the demand for oil and gas would continue to increase; and that the price for oil and gas would continue to rise. When the recession — U.S. and worldwide — lead to a *surplus* (rather than a shortage!) of oil and gas, and a *decrease* in prices, recession not only came to the industry, but the estimated return from projects undertaken based on these justifiable but erroneous assumptions nose-dived or disappeared as well.

As discussed in § **1.03(c)[2][b]**, PO and ROI calculations are not the most desirable methods of evaluating project profitability. However, such analyses do assist in evaluating Uncertainty Exposure, by showing how long your investment will be exposed to unexpected changes, and the sensitivity of the analyses to errors in estimating costs, overestimating return, etc. Likewise, a "Sensitivity Analysis" will show just how badly a proposed venture would be affected by wide swings in certain difficult-to-estimate variables.[79]

Where probabilities are difficult to estimate — such as the likelihood of an "on-trend" frontier wildcat well proving productive and the amount of reserves which would be established by the drilling of such a well — a "Probability of Success" evaluation may prove helpful. Such a review involves calculating what probability of success is necessary to achieve your minimum return objective. For instance, if a 1% success ratio on exploratory wells within

---

[78] This analysis is referred to as a "Cost of Failure" evaluation, and can be arrived at without determining whether or not the project as a *whole* should be undertaken.

[79] For instance, it may prove difficult to analyze just what may happen to oil and gas prices over the short term. However, if at today's cost levels and deal terms, prices could be cut in half and the project would still be viable, price can be seen as a less sensitive area.

the area would give the necessary Internal Rate of Return, undertaking the project might be an intelligent move. On the other hand, if to achieve the necessary IRR it was necessary that 90% of the exploratory wells be highly productive, a "go forward" decision is obviously an unrealistically optimistic one.

Insofar as Risk Analysis is concerned, one common approach is to make an *"Expected Value"* calculation. This analysis takes into account the probability that a possible result will occur by multiplying the probability of its occurrence times the presumed results if the event does occur.

When discussing a proposed exploratory prospect, you will frequently hear that the well has a "10% chance of success." This suggests that there is a 10% chance that the well will hit and also establish the anticipated reserves, and a 90% chance that the well will be a dry hole. In actuality, there are a variety of possible occurrences. The well may result in a dry hole; however, it is also possible that the well will be productive, but marginal or merely "commercial" (profitable, but not to the degree originally anticipated). There is also the possibility that the well will achieve the results projected by the geologist, or may even exceed the geologist's expectations.

An Expected Value analysis involves making a Present Worth calculation for each of these possible occurrences and then applying a "probability" factor to determine a risk-adjusted return. The sum of these risk-adjusted returns is then arrived at to determine the Expected Value of the project.

A sample Expected Value calculation is shown at **Figure 15**.

For any analysis to be meaningful, the assumptions on which it is based must be realistic. For instance, many companies "improve" projected (though no actual) return by assuming hefty and unlimited annual increases in the price of oil; therefore, their projected return appears much higher than the analysis prepared by a company

Event	Present Worth If Occurs	Probability of Occurrence	Risk-Adjusted PW
Dry Hole	[$ 400,000]	50%	[$200,000]
Marginal Well	[200,000]	10%	[20,000]
Commercial Well	400,000	20%	80,000
Excellent Well	4,000,000	15%	600,000
Exceptional Well	8,000,000	5%	400,000
			$860,000

EXPECTED VALUE = $860,000

Figure 15: Expected Value Example

which is more realistic in its revenue assumptions.[80]

Return analysis, to be useful, must be based on realistic assumptions. The importance of selecting a realistic discount rate in making Time Value calculations was emphasized in **§ 1.03(c)[2][a]**. Similarly, if revenue projections are too high (or too low), costs are understated (or overstated), or well "success ratios" are too optimistic (or pessimistic), the resulting "evaluation" will be useless as a decision-making tool.

§ 1.03(c)[4] Reaching a Decision. The analytical approaches just discussed represent a considerable oversimplification of the action by the petroleum economist in recommending to management what steps should be taken in deciding whether or not a project should be undertaken, and what type of deal structure would prove most desirable for the company. For instance, we have not discussed the "Decision Tree" approach, which involves plotting various possible alternatives in analyzing their expected outcomes. (A sample Decision Tree chart is illustrated in **Figure 16**.) However, you should now have some "gut feeling" about arriving at a decision on how a deal can best be structured from the standpoint of your company.

How do we put all this information together in actually reaching a decision? First, Internal Rate of Return should be used as *accept/ reject* criteria. In other words, in the absence of overriding non-economic factors, a project which cannot reasonably be expected to attain your company's minimum return objectives should not be

---

[80] The first company's projections show an even more dramatic "profit" if no corresponding increase in operating costs is projected.

Event	PW	Prob-ability	Risk Adjusted PW
Do Nothing	-0-	-	-0-
Drill:			
(a) Dry Hole	[100,000]	20%	[20,000]
(b) Productive	400,000	80%	320,000
(c) EXPECTED VALUE			300,000
Farmout:			
(a) Dry Hole	-0-	20%	-0-
(b) Productive	100,000	80%	80,000
(c) EXPECTED VALUE			80,000

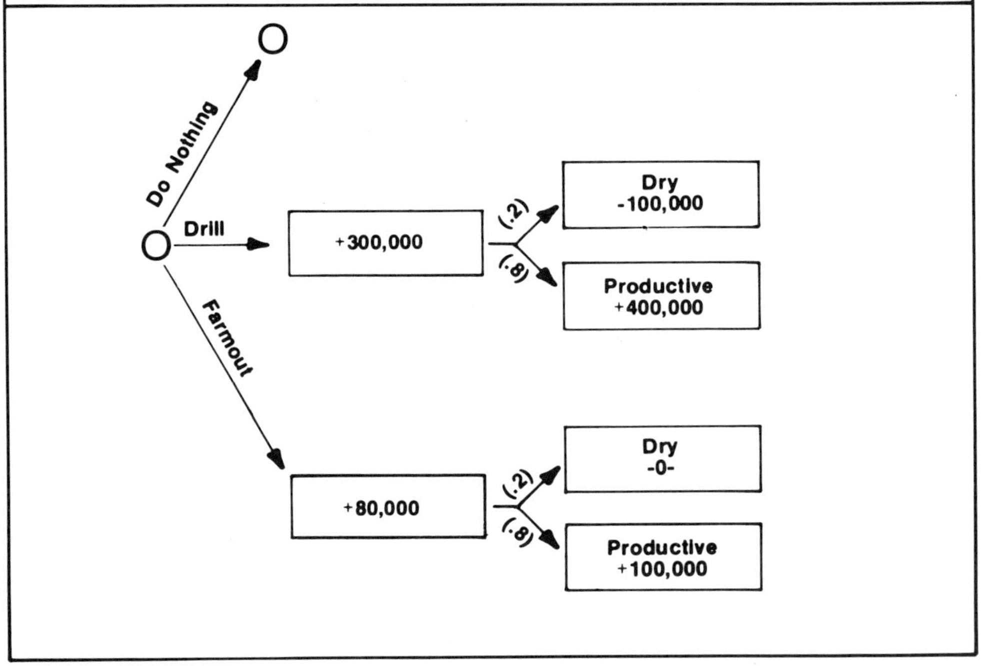

Figure 16: Sample Decision Tree

considered for acceptance.[81]

Due to the Risk and Uncertainty factors discussed in § **1.03(c)[3]**, no single project analysis will prove totally accurate. Such analyses, when properly applied to a number of prospects, customarily give results which are surprisingly valuable. And a continuing failure to analyze probable project profitability will prove disastrous to your company in the long run. (See § **1.03(c)[1]**.)

(Remember, of course, that these analyses will have been made for *each deal alternative* available to your company, and for *each different method* in which the deal selected — Farmout, Farmin, Support, etc. — might be structured. For instance, certain prospects that, when adjusted for Risk do not justify initial drilling for your own account, may make tremendous sense if the initial test well can be farmed out, and you will be participating only in the development drilling if the prospect proves productive.)

While the IRR analysis is useful in determining whether or not to even consider undertaking a project, the inherent problems in the IRR evaluation (see § **1.03(c)[2][b]**) do not make it the best criteria to apply in *ranking projects*. Here, PW/PV and DROI rankings provide much better guidance. Finally, other factors will be applied — such as upside potential, chances of establishing production in other zones or of improving return through future improvements in technology, downside risk, the length of time which the project will tie up your capital, the immediacy or deferred nature of cash flow, the impact of the project on other departments of the company, and management preferences, prejudices and predelictions — to determine which projects, under which deal structures, should actually be undertaken

---

[81] If you feel that the assumptions on which the analyses were made were unrealistically conservative, or if future price increases or improvements in technology which were not taken into account in the analyses are, in your opinion, most likely to occur or if the project has a tremendous "upside" potential, such speculative factors might lead to the acceptance of a project which, as analyzed, does not meet your company's IRR objectives. However, the acceptance of very many such projects tends to have a detrimental effect on job security unless your crystal ball is working overtime!

with the capital available.

Obviously, you are not always going to have your own way. You might prefer to farmout a given prospect, while your potential Farmee prefers that you join in the well. You may want to secure information merely through making a Support Contribution; the potential drilling party may be unwilling to let you share in that information unless you farmout. Here, negotiations begin which may result in your prevailing in your own desires, your giving in to the other party, a compromise solution being reached, or the deal falling through. However, without a proper analysis of the economics of the project, it is impossible for you to intelligently analyze how the deal should best be structured from the standpoint of your company, and whether or not a suggested compromise should be given serious consideration.

§ 1.03(d) Other Factors. In determining the best manner in which a deal can be structured, factors other than economic and tax considerations must be taken into account. For instance, what is your management's attitude toward this type of prospect? When are your leases due to expire? When are the other party's leases due to expire? What budgetary limitations do you face? What are the budgetary restraints, or lacks thereof, on the other party? What is management's general attitude toward farming in, farming out, or going it alone? Towards sharing well information? How much do you want to be a "partner" with the other party? Are there problems of rig availability — or a surplus of rigs — in the area; and if so, does the other party have access to rigs, or is willing to accept your proposal if his rig can be utilized?

It is important not to become locked into a position as to what management might or might not do based upon a pure economic analysis, when such an analysis does not take into account the practicalities of the particular situation with which you are dealing. To see a deal blow apart because management insists on a call on production when your leases are expiring and you have already been top leased is not the mark of a good explorationist. To insist on super tough Farmout terms in an attempt to secure additional testing of an area by another party where you have already drilled ten straight dry holes is also "Risk Free," because the deal is not going to be made. On the other hand, if it is the other party's leases which are expiring, that he has been top leased, and that his management has pulled the plug on further exploratory drilling in the area, you can insist upon much more favorable Farmin terms than you might normally hope to receive.[82]

§ 1.04 Guidelines for Selection Among Deal and Structuring Alternatives.

§ 1.04(a) In General. Throughout these volumes, we have and will emphasize the importance of tailoring the deal to the facts and circumstances of the particular prospect. Emphasis is also placed on your position, and the position of the other parties with whom you are dealing — vis-á-vis that prospect. Similarly, this portion of the book has only discussed the *types* of deals that are available to you. We'll now continue to discuss various ways in which a type of deal can be structured to make that deal more advantageous to your company (and often, at the same time, more acceptable to the other party), and how to avoid pitfalls which experience has shown tend to lurk in various types of arrangements. However, certain basic guidelines can be given which may be used as points of departure in trying to determine what type of arrangement might prove most beneficial:

1. If you consider your acreage quite valuable, but the

[82] Even under these circumstances, going for "overkill" is not a smart move. Someday, the tables may (and probably will) turn — it's a small industry, and memories are long. Furthermore, too tough a deal will often leave you with non-performance by the other party, expired leases, and a frequently useless "damages" remedy. Negotiating within a "zone of reasonableness," even when you have the stronger (or even an overwhelming) bargaining position, usually makes more business sense in the long run, particularly in the oil and gas industry.

acreage is sufficiently "rank" that current drilling on your part seems inadvisable, *giving a Support Contribution* to encourage another party to drill will give you valuable information at relatively little cost to your company. To Farmout a portion of your interest will be less desirable, since you would be decreasing your acreage position. However, if your leases are likely to expire unless drilling can be commenced in the area in the near future, or if you can Farmout a reltively small portion of your acreage for information that would prove extremely valuable, a Farmout is another alternative.

2. If management would be willing to accept the risks of drilling under the right economic circumstances, but economic analyses show that the return, when adjusted for risks, is unacceptable, the *acceptance of a Support Contribution* will change those economics. On the other hand, if economics are acceptable even without the support, and if information is at a premium in the area and a significant amount of open acreage (or leases that might be acquired by Farmout) is available, it may prove wise to decline the support to keep the competitive informational edge.[83]

3. If you have leases located in an area which is relatively "proven" as to certain zones, but has considerable untested potential in the deeper zones, and time is starting to run out on your leases, several solutions are available to you. First, you can undertake a program of *drilling to the presently proven zones.* If the economics of

---

[83] However, don't forget to take into account how difficult it often proves to keep information "confidential." (See § **2.06(c)**.)

such drilling do not meet management's return criteria, you can attempt to renew your leases. Similarly, management may determine that the need to maintain the deep rights more than counter-balances the disappointing nature of the return as to the proven zones. However, another alternative is to *Farmout as to the proven zones, reserving the deep rights.* (You may even want to consider making a "sweetheart" deal under the Farmout to insure a rapid development of the area and thus the extension of the deep rights under the vast majority [rather than only a portion] of your leases.)

4. If a prospect exceeds your normal return expectations, *securing a Farmin* from as many parties as possible is in your best interests, subject to any budgetary problems that such a policy might create. If your return projections are negative, attempt to farmout. (This is particularly useful when the return projections are negative only after the adjustment for the risks of *initial drilling.*) In an "Even Steven" situation, joining in the well, rather than accepting more or less than your normal share of costs and risks, may prove the most desirable solution. (This may also be the *only* solution, even on a high return prospect, where the other parties are unwilling to farmout and drilling on your own acreage would unnecessarily prove up the acreage of the other parties, or cannot be intelligently undertaken without their joinder.)

5. If you would like to join in a well, but budgetary constraints make full participation impossible, consider

a combined arrangement where you *join as to a portion of your interest and farmout as to the balance.*[84] Similarly, budgetary constraints may require you to permit other persons to join in what you consider a highly-profitable well where you would otherwise seek to secure a Farmin.

6. In a "frontier" area, where you have an acreage position throughout large portions of the area, a *Seismic Option*, with you as the *Optionor*, is an extremely useful method of securing additional information, and, hopefully, encouraging testing of the area. If your exploration department feels that the area has considerable potential, the Seismic Option, with you as *Optionee*, is also an extremely useful tool. Similarly, if a Lessor who owns choice acreage over an extensive area which is "wildcat" in nature is insisting on too tough a deal, a Seismic Option may be the solution.[85]

To summarize:

1. **Drill "heads up"** when:
   a. You have *sufficient budget*; and
   b. The *following don't seem important* (or desirable):
      1] *Sharing or reducing cost or risk* (Farmout or accepting Support would have been better);
      2] *Participating over a larger area* (Farmin or Working Interest Area would have been better);

---

[84] Another solution is to "promote out" that interest in the well which current budgetary restraints will not permit you to handle. This will also improve the economics of the deal from your standpoint as to your retained interest, since you will be receiving a disproportionately greater share of well revenues than that portion of well costs which you are bearing.

[85] As to the Seismic Option, see § **1.02(e)** and **Chapter Five**.

2. **Give Support** when:
   a. You *don't want to share or reduce cost or risk* (Joint Operations, Farmout, or accepting Support would have been better); and
   b. *You don't want to (or can't) increase your own acreage position* (Farmin would have been better); and
   c. *You need further information or don't want to "prove up" someone else's acreage* ("own account" drilling would otherwise have been better).

3. **Accept Support** when:
   a. *Economics are unacceptable;* or
   b. You are under *budgetary restraints;*
   c. You *don't want to increase or reduce your average position, participate over a larger area, or spread costs.*

4. **"Promote" or Farmout** when:
   a. You are under *budgetary restraints;* or
   b. You are willing to *reduce your acreage position and* ***aggregate*** *profit opportunity* ***to shift costs and risks and improve your return on investment.***

5. **Farmin** when:
   a. *Budget isn't a problem;* and
   b. You are willing to accept *greater costs and risks for an increased acreage position* and *a chance for greater* ***aggregate*** *profits.*

6. Form a **Single-Well JOA** when:
   a. *Economics are acceptable;* and
   b. You *don't want to/can't increase or spread your acreage position or exposure.*

7. Form a **Multi-well Working Interest Area** when:
   a. *Economics are acceptable over the entire area;* and
   b. You want to *spread your exposure over a broader area;* and
   c. You *don't want to/can't increase or reduce your aggregate acreage position or cost/risk exposure.*

8. Couple forming a **Working Interest Area with a Farmin** when:
   a. You want to *spread your exposure over a broader area;* and
   b. You are willing to accept *greater costs and risks for an increased acreage position and a chance for greater aggregate profits.*

9. Couple forming a **Working Interest Area with a Farmout** when:
   a. You want to *spread your exposure over a broader area;* and
   b. Are under *budgetary restraints;* or
   c. Are willing to *reduce your acreage position and aggregate profit opportunity to shift costs and risks and improve your return on investment.*

10. **"Frontier Areas":**
   a. **Grant a Seismic Option** when:
      1] You wish to *have the area evaluated and tested;* and
      2] *Budget is a problem;* or
      3] You are willing to *reduce your acreage position and aggregate profit opportunity to shift cost and risk.*
   b. **Accept a Seismic Option** when:
      1] *Budget isn't a problem;* and

2] You wish to *acquire or increase an acreage position and you are willing to accept increased costs and risks* ***if the area appears promising.***

c. Enter into an **Exploration Agreement** when:

1] You want to participate in an *evaluation, testing an development of the area;* and

2] You are willing to *pool your interests* within the area.

For a graphical illustration of the impact of various factors on "deal choice," see **Figure 17**.

§ 1.04(b) A Sample Case. The following illustration shows an application of the points discussed in this chapter.

Let's assume that your company has recently completed a seismic program in an area; the seismic indicates the presence of an anomaly which could be tested by drilling a 14,500 foot Red River test. The structure, as indicated by your seismic, and the ownership of leases within the area, is reflected on the Lease Ownership Plat shown at **Figure 18**.

The Red River formation historically produces only from structures. Similar tests have reflected a 35% chance of "success" for the initial drilling. The nearest Red River production is approximately 40 miles to the northwest.

You estimate that the necessary test well would run $1,060,000 for a dry hole, and $1,264,000 if the well is productive. You further estimate that the well, if productive, should establish recoverable reserves of 800,000 barrels of oil, to be recovered over a 15-year period. Oil is currently selling in the area for $28 a barrel. Estimated daily production would be 460 barrels of oil per day, and monthly operating costs should run approximately $1,000 a month. Your net revenue interest under your lease is 80%.

Spacing for similar prospects has been on a 160-acre (quarter-section) pattern.

Let's further assume that the other working interest owners in the

Budget a Problem:

1. *Accept Support* (some budget available; economics questionable or don't want to reduce acreage position)
2. *Farmout or Promote* (budget non-existent)
3. *Combine Joinder with Partial Farmout* (some budget available; economics acceptable)
4. *Form Working Interest Area* (to spread budget over more wells, or where a "partner" needed for complete development of prospect)

Need Further Information:

1. *Give Support* (don't want a "partner")
2. *Farmout or Give Drilling Option* (are willing to accept a "partner")
3. *Give Seismic Option or Execute Exploration Agreement* ("frontier" area)

Economics a Problem:

1. *Accept Support:*
   a. *DHC*: only *risk-adjusted* economics poor
   b. *BHC*: economics poor even without risk adjustment
2. *Farmout or Promote* (economics poor even with support)
3. *Form Working Interest Area* (spreading risk a key objective)

Economics Acceptable to Spectacular:

1. *Accept Support* (don't want more acreage or "partners")
2. *Join* (economics acceptable but not spectacular or want to spread exposure)
3. *Combine Support and/or Promotion with Farmins from Others* (want to increase acreage position without increasing cost/risk)
4. *Farmin or Accept Drilling Options* (economics good to spectacular)
5. *Accept Seismic Option or Execute Exploration Agreement* ("frontier" area)

Figure 17: Impact of Various Factors on Deal Decision

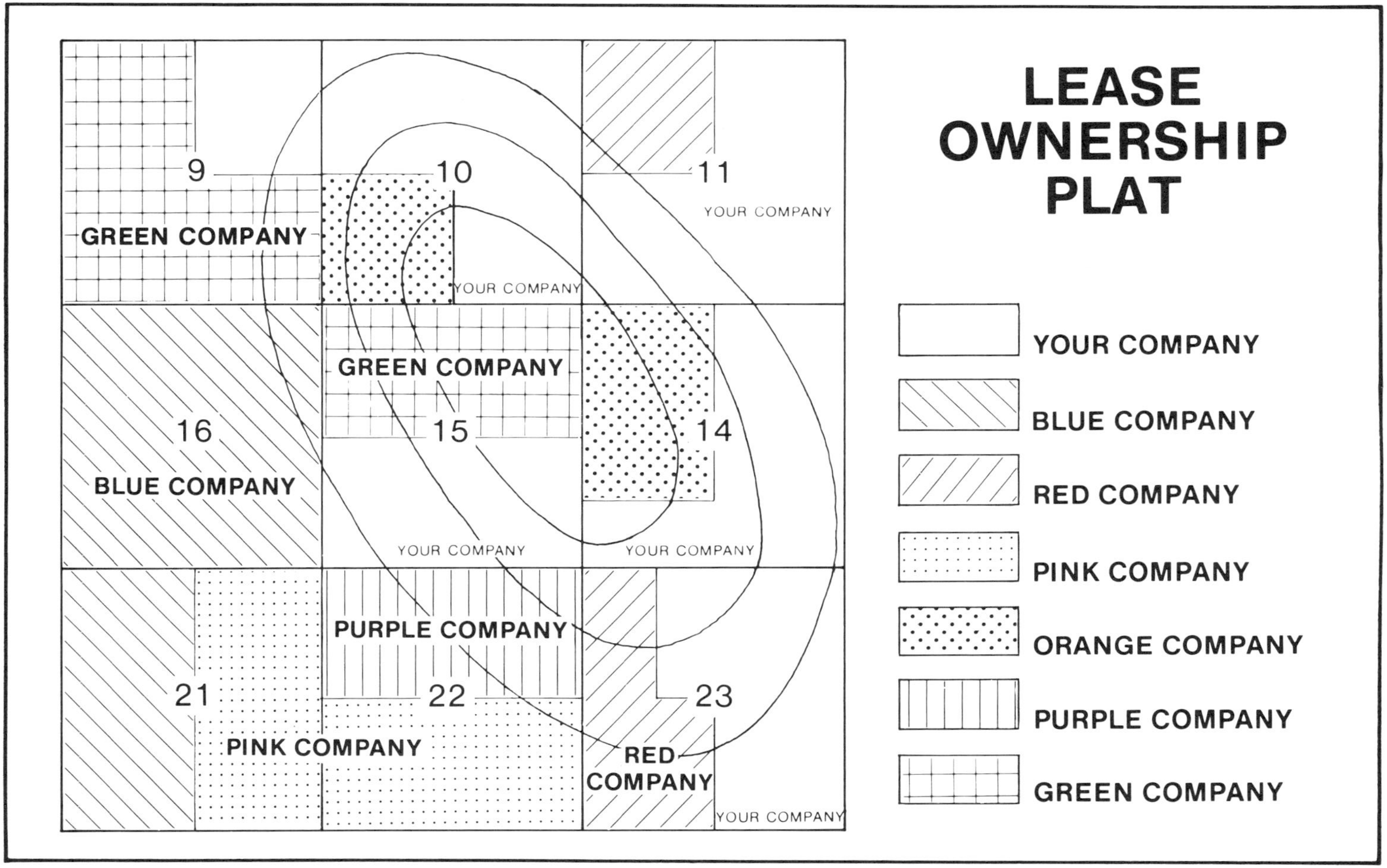

Figure 18: Lease Ownership Plat

area do not have access to seismic information, and that you do not intend to show them your information until a deal has been negotiated (if a deal can be made).[86]

A first question would be whether or not you would be willing to drill a test well "heads up" in this area at an appropriate location where you owned 100% of the Drilling and Spacing Unit. Here the answer is probably *yes*. Assuming that the well is productive, Payout should be achieved in approximately four years.[87] Your Return on Investment would be approximately 12.41 to 1, prior to adjustment for risk.[88] Even after adjustment for risk, the risk-adjusted ROI is still over four to one; and Payout, on a risk-adjusted basis, would still be achieved in less than 12 months.[89]

Also, you have not allowed for any future price increases

---

[86] This may be an unrealistic assumption, as negotiation of such deals often involves a "Show and Tell" as to any available seismic information.

[87] Assuming a 30-day month, and disregarding the Windfall Profit Tax and any decline curve, monthly production to your account equals 11,040 barrels of oil per month (460 BOPD x .80 x 30). Monthly net revenue to the working interest would be $308,120 (11,040 BO/M. x $28 = $309,120, less 1,000). This figure, divided into the estimated productive well costs, gives you Payout in 4.1 months ($1,264,000 ÷ $308,120).

[88] The cumulative cash flow over the life of the well is $17,920,000 (800,000 barrels x .80 x $28 a barrel). Maximum cash outlay is $1,440,000 (15 years production x $12,000 a year operating costs = $180,000, + $1,264,000 productive well costs). Cumulative cash flow (17,920,000) divided by maximum cash outlay ($1,444,000) = a 12.41 ROI.

Notice that by a little game playing this ROI can be "improved," even though the well economics have not changed. Thus, if you divided *net revenue* over the life of the well (i.e., the $17,920,000 *less the $180,000 lifetime operating costs*) by the *well cost alone* (1,264,000), you arrive at a 14 to 1 ROI!

[89] Multiplying the 12.41 ROI by the 35% success ratio gives a risk-adjusted ROI of 4.34. Multiplying the 4.1 month payout figure by the 35% success ratio results in an 11.7 month risk-adjusted payout.

This does not mean that any given well, even assuming the accuracy of your cost and revenue assumptions, would payout in approximately 12 months or return approximately 4 to 1 on your investment. What it does mean is that if three of these prospects were drilled, with a single well on each, and two proved non-productive while one attained the assumed results, the *overall* Return on Investment and Payout should equal these figures.

although you have also not upwardly adjusted future costs), and you have further ignored the improvement on economics that the lesser risks of drilling on development locations would provide.

(Yes, I *know* that I said petroleum economics are *always* computed on an after-tax basis, with costs and revenues adjusted for time. However, to avoid complicating this example, tax and Time Value considerations have been ignored.)

So, drilling a well "heads up" on this prospect would be one viable solution. However, there may be better ways to process this acreage.

The kind of deal you can negotiate with other parties will vary, as indicated in **§ 1.03(d)**. This will include such considerations as your management's attitude toward drilling (aggressive or non-aggressive); the similar attitudes of the managements of the other parties; your lease expiration dates versus the other parties' lease expiration dates; your budgetary limitations versus their budget limitations; and your management's attitude toward this type of prospect versus that of their management. The likelihood of "seismic slippage" on prospects of this type must also be taken into account.

Assuming that none of the parties will overly benefit or will be overly burdened by such considerations, what would be your orders of preference as to structuring an appropriate deal and/or drilling for your own account?

By showing your seismic (which we said we did not want to do), you could almost certainly get Green or Orange to accept a Farmout of your acreage. However, since this appears to be an attractive prospect — and we're assuming no particular management disinclinations for such a test — this would be a last choice. You could also have a well drilled further to evaluate your acreage by making a Dry Hole or Bottom Hole Contribution to Green or Orange; however, by doing so, you have wrecked any likelihood of securing a Farmout from Green and/or Orange, and would seriously compromise your ability to get these parties to agree to the formation of a Working Interest Area.

Another way in which a well could be secured in the area is to sell an interest in your acreage to a third party on a "promoted" basis, under which he would pay a disproportionate part of well costs. If you were subject to severe budgetary restraints, this might be a solution. (See § **1.04(a)**. However, we have assumed that you are not; and since this is a desirable prospect, diluting your interest — even on a promoted basis — would not seem the most desirable method of structuring this deal.

A better solution than those considered above would be to form a Working Interest Area, under which the interests of the parties would be pooled; and to have expenses and production shared in proportion to the lease acres of each party within the Contract Area.[90] While your interest would dilute, you would be sharing in a greater number of wells, including wells located on acreage which may be better positioned on the structure than your own leases. If this were the decision, the further determination of what acreage should be included within the Working Interest Area would depend upon the likelihood of productivity and/or profitability of wells drilled on the flank of the structure, and how comfortable you were with the accuracy of your seismic interpretation, i.e., the likelihood of "seismic slippage."

An equally-attractive solution might be simply to drill 100% for your own account on your own acreage. In doing so, you would eliminate the need for partners; however, you would have the exposure to a greater degree of risks — both of the "dry hole" and "seismic slippage" variety — without the diversification that the formation of a Working Interest Area would have provided.

If the decision is made to drill "heads up," another decision must be made on whether or not to accept support from the other parties. Without support, you are proving up the acreage of the other parties

---

[90] Another solution would be to base the Sharing Arrangement on the *relative value* of each party's acreage, based on position of the structure. However, this is difficult to evaluate; even more difficult to negotiate; and might not work to your best interests, vis-à-vis Green and Orange. See **Chapter Four**.

without their sharing in risks. By accepting support, you have reduced your own risk, even if the support is "dry hole" in nature only. However, you have also given up any advantage of the exclusivity of your information, and have wrecked your opportunity of negotiating Farmins from the other parties.

A final solution remains which would appear to be preferable as to this particular prospect. This would be to negotiate Farmins from the other parties, coupled with the formation of a Working Interest Area comprising a reasonable amount of acreage within the prospect contours. Under this arrangement, your company would pay 100% of the costs of the initial well drilled on the area; however, by doing so, you would earn an interest in the leases owned by the other parties, and thus would increase your interest in the Working Interest Area. The combined Working Interest Area - Farmin arrangement, as contrasted with negotiating separate Farmins with each of the parties owning leases in the area, would also permit you to earn an interest throughout the area without the necessity of drilling multiple earning wells.

Assuming that this last solution is the one you decided to attempt to negotiate, and assuming that those negotiations were successfully conducted, a final determination would be the size of the Working Interest Area. One obvious solution would be to include only the acreage on the top of the structure owned by you, Orange and Green, e.g., the South Half of Section 10, the Southwest Quarter of Section 11, all of Section 15, and the East Half of Section 14. This would permit you to retain the balance of your acreage in Sections 10, 11, and 14 for own account drilling if the seismic should slip to the north or to the east. Another possibility would be to form a larger unit area, excluding as much of your acreage in Sections 11, 14, and 23 as possible, but picking up a substantial portion of the acreage of Red, Purple, and Blue. The inclusion of the Red, Purple, and Blue acreage would prevent your interest from being diluted, since you would be earning a significant interest in that acreage by drilling the initial test well; and would also protect you against "seismic slippage" to the south or west. It would also be beneficial if you were interested in the possibility of improved production on the flanks of the structure — a real possibility in Red River production in the Powder River Basin — or in the possibility of *stratigraphic* plays in other formations.[91]

§ 1.04(c) Conclusion. The principles discussed in this chapter are essential to understanding the "whys" and "hows" of exploration deal structure; to a determination of the basic types of exploration deals available to you; and how economic, tax, and other considerations influence the selection of the proper deal type. The balance of this book and subsequent Volumes will deal with further considerations for properly structuring your deal after a basic deal type has been agreed upon, and to inform you of pitfalls to avoid under these various types of arrangements.

---

[91] Note that you have maintained a relatively high interest in the larger unit area by bringing in the Red, Purple, and Blue acreage at the expense of Green and Orange. Accordingly, your ability to negotiate such an arrangement with Green and Orange would be dependent upon there being no "Show and Tell" of your seismic to those parties, or their similar concerns of "seismic slippage," flank production, etc.

Chapter Two

SUPPORT AGREEMENTS

§ 2.01 Basic Concepts

§ 2.01(a) The Whats and Whys of Support Arrangements. Two parties own acreage in a prospect area. One is considering the drilling of a well; however, the cost of that well, or the potential return in relation to well cost, seems out of line considering the degree of risk involved. The other party would also be willing to drill on his own acreage — if he had more information.

Neither party considers that the transfer of initial drilling risk to the other is sufficient motivation for losing control of his acreage position. Accordingly, a Farmout is not the solution. On the other hand, neither particularly wants to be a "partner" in wells on the other party's acreage, nor to have the other party as a partner in wells on his acreage.

The needs of both parties can be satisfied by one lease owner entering into an arrangement to "support" the other party's well through *a contribution of cash* or something else of value (see **§ 2.03(a)**) in return *for information*. The Supported Party—the party receiving the contribution—reduces the cost of a dry hole (and thus the dry hole risk), or, in some instances, the cost of the well whether productive or dry. The party making the contribution—the Supporting Party—secures the information required to evaluate the desirability of drilling on his own acreage at a fraction of the expenditure that would have been involved in the drilling of an initial test well.

The *Support Agreement* is thus a contractual arrangement under which the Supported (or Drilling) Party receives a contribution of something of value from the Supporting Party who owns acreage whose value could be better evaluated based on information gained by the drilling of a well on acreage owned by the Supported Party. The Supported Party earns the contribution by performing specified drilling and testing obligations, and by sharing the results of that testing with the Supporting Party. (See § **2.04** *et seq.*)

A sample Support Agreement is contained in **Volume Four**.

§ 2.01(b) The Pros and Cons of Accepting Support. The advantages of the Support Arrangement to both parties are discussed in § **1.02(b)**. (See also § **1.04(a)** as to circumstances under which you might want to give, or accept, support.)

At first blush, it might appear that the Drilling Party is always in a "no-lose" situation if he accepts proffered support. However, it's in fact a serious decision for a Drilling Party as to whether he should accept support rather than insisting on a Farmout, the formation of a Working Interest Area, or a combination of both. Due to the value of both acreage and information, it is getting more difficult to get some companies to accept support (and thus share information) without permitting them to also acquire additional acreage. Thus, while the Drilling Party's risk of drilling a dry hole is obviously reduced by accepting support, he is also giving up his bargaining power to try to improve his acreage position. This is a trade that a number of companies are becoming more hesitant to make.

Also, the Drilling Party must always balance the value of the anticipated support versus the loss of the competitive edge that comes from having exclusive information. While the information will eventually become public, this release may not be in sufficient time to permit other companies interested in the area to make intelligent decisions on whether to lease or reject existing open acreage, to Farmout rather than drill at their own cost or insist on joint participation, etc.

(Of course, there is always a question of just exactly how "exclusive" such information will remain, even where the well is being drilled as a "tight hole." (See § **2.06(c)**.)

§ 2.01(c) Types of Support Arrangements. Support Agreements are customarily classified as *"Dry Hole Contributions"* or *"Bottom Hole Contributions,"* and as *"Cash Contributions"* or *"Acreage Contributions."*

Support Agreements are called Dry Hole Contributions ("DHCs") if the support is to be earned only if the earning well is non-productive; Bottom Hole Contributions ("BHCs") are arrangements where the contribution will be made whether the well is productive or dry.

Cash Contributions involve support which is to be made by a cash payment. Acreage Contributions support a well by the assignment of an interest in acreage—frequently, a small interest owned by the Supporting Party in, or immediately adjacent to, the Drilling Party's well location.[1]

Problems frequently arise in determining whether a given earning well is or is not a "dry hole" for purposes of earning a Dry Hole Contribution. There is also no reason that a *"Marginal Well Contribution"* could not surface as a third category, midway between the Dry Hole and the Bottom Hole Contribution. See **§ 2.04(d)**.

Similarly, the division of Support Agreements between Cash Contributions and Acreage Contributions suggests that a contribution *must* be limited to cash or acreage. However, the creative explorationist will find other "things of value" that his company may contribute to encourage the drilling of a well—and the sharing of the resulting information. (See **§ 2.03(a)**.)

§ 2.02 Points to be Covered in Support Negotiations

§ 2.02(a) In General. In negotiating a Support arrangement, and drafting the resulting Support Agreement, the parties will want to consider the following points:

---

1 The fact that the Supporting Party retains no interest in the acreage assigned distinguishes the Acreage Contribution from a Farmout.

1. **Nature and Extent of the Support.**
What *kind* of support will the Supporting Party offer? What is the *extent* or amount of that support? Should any incentives or *"sweeteners"* be offered to encourage the Drilling Party to do more than the minimum required to earn the support? (See § **2.03** *et seq.*)
2. **Earning Requirements.** What requirements concerning: (i) the *earning well* (location, commencement and completion dates, depth, etc.), and the (ii) *testing requirements* (tests to be conducted, samples to be taken, information to be provided, opportunities for notification and access, etc.), including provisions to protect the Supporting Party against inadequate information, and the Supported Party against overly burdensome (or even "impossible") requirements, should be imposed as conditions of earning the support? (See § **2.04** *et seq.*)
3. **Miscellaneous Provisions.** Are there to be any unusual requirements concerning those areas usually looked at as "boiler plate," such as notice, assignability, confidentiality of information, time for acceptance, etc? (See § **2.06** *et seq.*)

§ 2.02(b) Inadequacy of Coverage. The typical Support Agreements in use today are based on forms developed many years ago when wells were shallow, involved much less cost and risk, and were far less apt to cover prospects with multiple objective horizons. To try to cover today's complicated drilling arrangements with these "Support Letters" is like trying to win the Indianapolis 500 with a Model "T" Ford—it was fine for its time, but it's now out of date. Yet that is the basis on which nearly all "modern" Support Arrangements are being negotiated and reduced to writing.

To a great extent, today's Farmout Agreement has kept pace with the times. Yet Support Agreements are still negotiated based on brief telephone conversations that cover only well specifics, the dollar amount of the support, and whether the contribution is to be "dry hole" or "bottom hole." The forms used to reflect the agreement are equally inadequate.

If a relatively inexpensive 4,000 foot test is involved, perhaps this is sufficient (or at least all the press of time will permit). Where a multi-million dollar rank wildcat is at stake, this cavalier treatment is foolhardy.

The balance of this Chapter will explore Support Agreements and various structuring alternatives for today's deeper tests.

§ 2.03 Nature and Extent of Contribution

§ 2.03(a) Nature of the Contribution. Support Contributions are most commonly made in the form of a cash contribution towards the cost of the well.[2] In some instances, the contribution will be in the form of acreage—an assignment of an interest in certain leases in which the Supporting Assignor, unlike a Farmor, retains no overriding royalty or similar interest in production.

The cash/acreage alternatives are so common that we often forget that the name of the game is to find *whatever* will motivate the Drilling Party to drill the well and share the information. And this doesn't have to be cash or acreage: it can be *anything of value* sufficient to induce the Drilling Party to drill—and share. Thus, anything that you can provide a potential Drilling Party to promote that drilling and information sharing should be considered as a method of accomplishing *your* goal.

The "something of value" given to the Drilling Party could include the right to a Farmout if he first drills *this* well (a Farmout or Drilling Option [see § **1.02(c)** and **Chapter Five**]), the contribution of seismic data you possess, or even a call on your production within a given area. It could be to provide technology, such as permitting your personnel to perform a proprietary completion technique on the Drilling Party's well. The main thing is: get the well drilled, and get access to the information.

---

[2] As to the tax treatment of such cash contributions, see § § **1.03(b)[4]** and **2.07**.

(Of course, deciding what *will* motivate the other party to make this deal may require innovative thinking on your part!)

§ 2.03(b) Extent of Support

§ 2.03(b)[1] In General. Obviously, whatever the nature of the support, you must also decide *how much* of that support to give.

Where an Acreage Contribution is involved, the support is usually limited to certain small interests owned by the Supporting Party in the earning well Drilling and Spacing Unit (so small that it isn't worth the effort to participate in the well or to farmout), or immediately adjacent to that Drilling and Spacing Unit.

(Remember—Acreage Contributions are hard to get.)

However, the usual question concerning the amount of support is how much *cash will it take* (and is it worth) to convince the Drilling Party to share the information. (See § **2.01(b)**.)

There is a great variance in industry practices concerning when to support and the amount of support to be given to a particular well. Certain companies use fixed formulas as to both when they will support a well and the amount of their support. Other companies are quite flexible, basing the decision of whether or not to support at all, and how much to support, on the contributing party's need for the information and the value of that information to the contributing party.

One approach used by some companies is to allocate some specified percentage (often up to 50%) of the estimated dry hole cost of the well to be supported among the parties owning the direct and diagonal offsetting Drilling and Spacing Units (including the Drilling Party, if he owns offsetting acreage), with this contribution being allocated among these parties on an acreage basis. Other approaches are far more elaborate, taking into account acreage even more distant from the well, with complicated formulas as to how much cost it is reasonable to "assess" to each more distant tier.

Traditionally, the amount of support is expressed as a specified amount *per foot drilled,* up to a specified *maximum dollar amount.* But see § **2.03(b)[3]**.

As to "sweeteners" to encourage deeper drilling, see § **2.03(b)[4]**.

§ 2.03(b)[2] Supporting a Test to a New Horizon. Special problems arise in determining the proper amount of support when a well is being drilled in an area that is "mature" in the upper horizons, but is untested—a "rank wildcat"—as to deeper zones. If the Drilling Party is drilling on such acreage, his risk is primarily limited to depths below the known existing horizons; and the Supporting Party will essentially gain new information only as to drilling below those known producing zones. As a result, many companies limit their support of such a well to the footage drilled *below the base of the known existing formations.* Thus, if $25 a foot would be your company's normal support contribution on an exploratory test, and a well projected to 19,500 feet was expected to encounter a known-producing horizon at 6,500 feet, the support would usually be limited to $25 per foot drilled *below 6,500 feet,* not to exceed $325,000 ($25 a foot times 13,000 feet [19,500 feet less 6,500 feet]).

This approach may be game playing. The Drilling Party either will or will not give up the information for the agreed support. But, no matter how you arrive at the amount of the support, unless the overall dollar figure reaches a certain amount, he's not going to play. Accordingly, and "formulas" to the contrary, we're still dealing with a one-on-one negotiated arrangement. Furthermore, the cost of drilling the well from the surface to the shallower formation will be significantly higher when the well has been engineered to a much greater total depth than if the well had been engineered to the depth necessary to reach the shallow formation alone. Thus, the question of whether or not to limit the amount of support on such a test is basically one of negotiation: how much does the Drilling Party need the support and how badly does the Supporting Party need the information?

For other problems concerning support of a deep test to a new horizon, see § **2.04(d)**.

§ 2.03(b)[3] Maximum Dollar Amount Limitations. We've indicated that the traditional way of calculating support is to allow a certain dollar amount per foot drilled, *not to exceed a specified maximum amount,* based on estimated well cost and/or estimated depth. Since in a truly "rank" exploratory area the likelihood of cost overruns and unexpected conditions are the biggest "risks" that a party drilling a well will face, this "traditional" solution may not be a sufficient one for today's drilling conditions.

The Supporting Party will gain valuable information which will assist him in drilling on his own acreage. He will know whether or not cost overruns are likely to occur, what drilling conditions may be encountered, and which solutions do and do not seem to work in overcoming those conditions. Accordingly, a Support Agreement which is based on a *percentage of cost actually incurred* would seem to be fair to both parties, *so long as* it is subject to some maximum cutoff point, e.g., 150% of the estimated dry hole cost, as shown by the *Authorization for Expenditure* ("AFE").

§ 2.03(b)[4] Incentive Arrangements. Often, a rank wildcat could have multiple potentially-productive horizons. The deepest possible zones may be ones to which the Drilling Party may be unwilling to *commit* to drill, although such zones may in fact be tested, depending upon drilling conditions encountered, etc.

Obviously, if the well is drilled only to relatively proven or relatively shallow horizons, you won't be interested in supporting the well. However, if the test is a "deep" one—although not as deep as you might optimally hope for—you may still be willing to support the well in return for the information. However, the Support Agreement should encourage drilling to the deepest possible zones.

Among the "sweeteners" you could add to the deal to encourage the deeper drilling are the following:

1. Dry Hole Support converted to Bottom Hole Support if

the deeper zones are drilled.

2. A greater dollar-per-foot support.[3]

3. A percentage of actual costs, with no dollar limit (other than some relatively high ("upset") limit, if the deeper zones are tested. (See § **2.03(b)[3]**.)

4. A Farmout Option if the deeper zones are drilled.

Similar "sweeteners" could be used to encourage extremely expensive testing.

(These are examples only; use your imagination to dream up other ideas.)

As to the use of incentive arrangements in Farmouts, see § § **3.03(b)[6][c]** and **3.04(b)[3][d]**.

§ 2.04 Earning Conditions

§ 2.04(a) Earning Well Requirements

§ 2.04(a)[1] In General. Provisions concerning the earning well under a Support Agreement are quite similar to the earning well provisions in a Farmout Agreement. Thus, the agreement must specify the well location, the minimum depth to which the well must be drilled to earn the support, the commencement date, and the completion date.

Most of these considerations will be discussed in connection with the Farmout Agreement. (See § **3.03(b)** *et seq.*) However, there are certain problems concerning the earning well which are unique to the Support Agreement, or create special problems in Support Arrangements, which will be discussed here.

§ 2.04(a)[2] Well Commencement and Completion Dates. The Support Agreement will specify the date by which the earning well must be commenced, and the date by which the well must be

[3] This could be limited to the deeper footage only, or could apply "top to bottom."

completed. As in the case of the Farmout Agreement, it is critical that there be no uncertainty as to exactly what constitutes the "commencement" and "completion" of the well.[4]

The Drilling Party must insure that each of these dates is one with which he can reasonably comply. On the other hand, if there are extensive delays before significant information can be delivered to the Supporting Party, or if the Supporting Party does not receive the information on a timely basis, the information may prove valueless; similarly, in most instances, the quicker the information is received, the greater its value to the Supporting Party.

As is the case in the Farmout Agreement, the Drilling Party must insure that rig availability and title problems will not prevent him from being able to commence the earning well by the specified commencement date. Conversely, from the standpoint of the Supporting Party, the earlier the date by which the well is commenced, the quicker he will start receiving information and the more valuable that information will be to him. Accordingly, a provision which may satisfy the needs of both parties is to provide for an initial commencement date which can be extended *so long as* the Drilling Party is acting with diligence and the failure to commence by the specified date is due to circumstances beyond the Drilling Party's control.

In a Farmout Agreement, the Drilling Party would frequently only be required to drill the earning well with diligence, and no final cutoff date might be provided.[5] However, in a support situation, if the information reaches the Supporting Party too late to permit him to take necessary action to extend or develop his leases, the information will be valueless to him. Accordingly, he may not be willing to accept a mere "drill with diligence" provision. On the other hand, from the standpoint of the Drilling Party, a fixed cutoff date

---

4 As to the definition of "commencement" and "completion," see § **3.03(b)[1]**.

5 See § **3.03(b)[1]**.

which did not adequately take into account the possibility of unexpected delays in drilling (which are to be expected in the initial testing of this type of a prospect) might make the promised support illusory. A possible solution is to provide that the Drilling Party will drill the earning well with all due diligence and complete the well by a specified date, again subject to extension if the inability to complete by such date is due to circumstances beyond the Drilling Party's control. However, to protect the Supporting Party against the information being delivered to him at such a late date that the information is valueless, a final, absolute cutoff date should be provided.

In many instances, the Supporting Party will have received valuable information throughout the course of drilling the well which permits him to take significant action concerning his leases even though the earning well cannot be completed by this final cutoff date. Information concerning intermediate, previously untested formations may be extremely valuable to the Supporting Party in evaluating his leases. However, under the typical Support Agreement, the fulfilling of the earning well conditions, including the completion date, is a black or white situation: if the well reaches total depth and is completed by the specified date, the contribution is earned; if it doesn't—no matter how valuable the interim information provided the Supporting Party, and how understandable the delays—no contribution need be made. For this reason, a modification of the customary black/white earning provision should be considered, which will provide for *partial support* in the event that significant information of value has already been received by the Supporting Party by the final cutoff date, even though the earning well cannot be completed by that date. See **§ 2.04(c)**.

§ 2.04(a)[3] Minimum Depth ("Contract") Provisions. Many of the problems concerning the minimum depth to which the earning well must be drilled in order to earn the support are common to both Support and Farmout Agreements. (See § **3.03(b)[4]** *et seq.)*

However, Support Agreements customarily have not contained the "escape" clauses which are customary in Farmout Agreements, since there is no *obligation* on the part of the Drilling Party under most Support Agreements to drill the earning well. (In other words, the well is merely a *condition* of the support, not a mandatory or "covenant" well.)[6] Accordingly, most Support Agreements also do not contain a "Substitute Well" provision.[7]

There is no reason that a Support Agreement should not contain a substitute well provision, *so long as* the substitute well can still be completed by the final cutoff date provided for the drilling of the earning well. (See § **2.04(a)[2]**.) Therefore, the Drilling Party may choose to request such a provision if expected subsurface conditions seem to warrant it. However, if such a provision is to be inserted, the Supporting Party should insist that the well be completed by the specified cutoff date if the full support is to be earned.

Just as problems arise concerning an inability to complete the earning well by the final cutoff date (See § **2.04(a)[2]**), a similar problem may arise as to whether or not the earning well can be drilled to the total *Contract Depth.* Accordingly, some consideration should be given to a partial earning in the event that significant horizons have been tested, and the inability of the Drilling Party to drill to the total Contract Depth is due to circumstances beyond his control, or other conditions that would still permit him to earn under a Farmout Agreement. See § **2.04(c)**.

§ 2.04(b) Testing and Information Requirements

§ 2.04(b)[1] In General. Unlike the Farmout, the only thing of value which the Supporting Party receives under a Support Arrangement is his right to information. If the tests are not adequate to provide

---

[6] As to excuse provisions in Farmout Agreements, see § **3.03(b)[4][c]**.

[7] As to substitute well provisions in Farmout Agreements, see § **3.03(b)[4][d]**.

meaningful information, the Supporting Party has been bilked. On the other hand, since none of the Supporting Party's money is at stake, he might like to see tests conducted which are totally unreasonable from a cost standpoint and from the standpoint of information which will also be of value to the Drilling Party. This is particularly true where velocity survey and other tests are involved which give "trend" information concerning adjacent acreage in which the Drilling Party may own no interest, and which will have little value in evaluating the earning well.[8]

Another "Catch 22" is that where "rank wildcat" acreage is involved—the situation where a Support Arrangement will be most useful—neither of the parties will know until the well is drilled exactly which formations should be tested and which tests should be run.

How are we going to resolve these problems?

§ 2.04(b)[2] Nature and Extent of the Testing. The Support Agreement should clearly specify exactly which tests are to be performed by the Drilling Party. From the standpoint of the Supporting Party, it would be desirable to require the performance of such tests, and the providing of such cores, cuttings, samples, logs and surveys, as the Supporting Party, in his absolute discretion, might request. Since the Supporting Party normally prepares the Support Letter, this is exactly what the Agreement, as submitted to the Drilling Party, will normally require.

If the Drilling Party has had prior dealings with the Supporting Party, he may feel comfortable with such an "absolute discretion" provision. However, it is dangerous from the standpoint of the Drilling Party to permit such unlimited discretion in the Supporting Party. Therefore various alternatives should be considered:

(1) *Limit testing to such tests as the Supporting Party, "as a*

---

[8] This is another reason that the Drilling Party may determine that accepting support may be more costly than it is worth.

prudent operator, shall reasonably require in accordance with customary operating practices in the industry." This provision has the advantage of giving considerable flexibility; however, it provides no objective (fixed) standard, and can lead to substantial disagreement—good faith or otherwise—between the parties.

(2) *Specify in the Agreement the exact number and nature of the tests to be performed.* This certainly provides an objective standard; however, as mentioned in § **2.04(b)[1]**, the exact formations to be tested, and the tests to be performed, may be quite difficult to determine except during the course of drilling the earning well.

(3) *Fix a specified number of Drillstem Tests and/or feet of core, or a dollar amount of testing, that can be required by the Supporting Party.* Where such testing is to be performed will be totally in the discretion of the Supporting Party. Nevertheless, in the event that additional tests are requested by the Supporting Party, they will be at his own cost, or the costs of such additional tests will be split between the parties in some pre-specified percentages. (Another variation would be to provide for an additional dollar per foot [or additional percent of actual costs] for each test required above the number fixed in the Agreement.) This arrangement has the advantage of permitting the Supporting Party to select those tests he most wishes to see performed, while placing some limit on "unreasonable" requirements.[9]

§ 2.04(b)[3] Information and Access. The mere conduct of the testing is not what the Support arrangement is all about from the

[9] However, if the Supporting Party's drillsite geologist is gung-ho, or the Supporting Party is somewhat lavish (at least in the eyes of the Drilling Party) with his own money in the conduct of testing, a limit of the cost sharing variety will not insure that testing costs do not greatly exceed what the Drilling Party had budgeted.

standpoint of the Supporting Party: it is also critical that the information that results from such testing be provided to the Supporting Party, and that it be provided in a timely fashion. Furthermore, it is not merely access to the Drilling Party's conclusions that is critical; an opportunity to be present while tests are being performed and otherwise to have access to the rig floor is vital, since the *interpretation* of the information provided by the testing is as critical as the availability of the information itself.

The Support Agreement should clearly specify how many copies of each log, and what quantity of cores, cuttings, samples and the like are to be provided the Supporting Party. (In this regard, as discussed in § **2.04(b)[2]**, it is critical that the coring and similar requirements not be excessive, particularly if similar information must be provided several Supporting Parties.)

It is also vital that the information be provided to the Supporting Party in a timely fashion. While the "access" provisions may assist in this regard, the Supporting Party should insure that he will receive the information on a timely basis, so that he can take prompt action in securing additional leases and otherwise be able to act promptly on the information provided. Likewise, the "access" provisions must be coupled with notification requirements so that the Supporting Party will know when tests are to be performed, or points in the drilling are to be reached, in the event he wishes to have his personnel present.

In many instances, the Drilling Party will perform tests which are not required by the contract, or will drill to depths greater than the *Contract Depth* specified in the Support Agreement.[10] Normally, the Supporting Party must be provided all information as to tests actually run, whether or not required by the contract, and irrespective of the depths. However, under certain circumstances, the Drilling Party may wish to provide that the Supporting Party have no right to receive this additional information (it is unlikely that

---

[10] See § **2.04(a)[3]**.

the Supporting Party will agree to this), or that the Supporting Party be required to make some additional contribution in order to receive the information.

The Support Agreement should also specify whether or not the information concerning the results of such tests must be provided the Drilling Party—if the Supporting Party takes over the well to perform his own tests (see § **2.04(b)[4]**).

The Drilling Party may often determine that it would be in his best interests to forego the support and go "tight hole" on the well. While this would violate the spirit of the Support Arrangement and forfeit any right to support, it would normally not be a breach of the wording of the typical Support Agreement. Accordingly, the Support Agreement should provide that, *as to tests actually run,* the Supporting Party must be provided such information. See § **2.06(a)**.

§ 2.04(b)[4] Testing by the Supporting Party. The Supporting Party may wish to add a provision to the Agreement, comparable to similar provisions in Farmout Agreements (see § **3.04(c)**), giving him the right to take over the well to perform his own tests which he cannot reasonably request of the Drilling Party. The use of such *Take-Over* rights to overcome stalemates as to the extent of the tests to be performed has been discussed in § **2.04(b)[2]**. Here, we are talking about tests which are clearly beyond the scope of the testing which can reasonably be required of the Drilling Party.

While such Take-Over provisions are quite common in Farmout Agreements, the Drilling Party may be far more reluctant to give the Supporting Party the right to take over his well, even on a temporary basis, in a non-Farmout situation. If such a provision is inserted in the Support Agreement, it should be coupled with adequate indemnification provisions; and the Drilling Party may wish to reserve the right to refuse to permit such a take-over if he feels that the tests to be performed might endanger the well. See § **2.06(b)**.

§ 2.04(b)[5] Substantial Compliance. Problems concerning the failure of a Drilling Party *literally to comply* with requirements concerning the well completion date, or drilling to Contract Depth, have been discussed in § § **2.04(a)[2]** and **2.04(a)[3]**. Such requirements become even more severe when a Drilling Party is prevented, due to circumstances beyond his control, from literally complying with each and every one of the tesing requirements, particularly the requirement of the running of a final log from the surface to the total depth drilled.

Just as some sort of partial support and/or substantial compliance provision may prove desirable in the event of a failure to attain total Contract Depth or a failure to complete the well by a final cutoff point, where a Supporting Party has received valuable information as a result of the testing—even though every test required by the Support Agreement could not be performed—some substantial compliance or partial support provision may be in order. See § **2.04(c)**.

§ 2.04(c) "Substantial Compliance" and Partial Earning Provisions. As mentioned in § **2.04(a)[2]**, the typical Support Agreement is an either/or (black/white) arrangement: either the Drilling Party strictly complies with each and every one of the conditions of earning, or no support whatsoever is to be given. While such an arrangement may have been reasonable in earlier, less complicated days, this type of earning arrangement is not responsive to the conditions encountered in the very type of "rank wildcat" exploratory drilling where Support Contributions are most called for.

In the past, where parties tended to honor the *spirit* of an Agreement, irrespective of the literal contract wording, such black/white earning provisions in the contract usually did not create problems in practice. However, recent tendencies toward disregarding the underlying spirit of an arrangement in favor of a literal interpretation of contract wording make the "all or none" nature of the typical Support Agreement more dangerous from the standpoint of the Drilling Party. (See the **Introduction** to this book.)

Under a Support arrangement, the Supporting Party is in the driver's seat. By the time the well has been drilled, and the Drilling Party has a right to demand the support, the Supporting Party has already received all of his information. It is highly unusual for a Support arrangement to provide for "Payments on Account" until the drilling of the earning well, and all testing requirements, have been completed. Accordingly, it is quite simple for the Supporting Party at that stage to raise some minor technicality and to withhold the support. The Drilling Party's only remedy is to sue to collect the support. Such suits are distasteful to most oil and gas operators and frequently do not provide a practical remedy. In addition, it may often prove impossible for the Drilling Party literally to comply with every minor detail of the contract earning provisions.

One solution would be to insert a *Force Majeure* provision in the contract, whereby the support would be earned if the inability of the Drilling Party to strictly comply with the provisions of the agreement were due to circumstances beyond his control. However, Force Majeure arrangements are inserted in a contract to excuse a party *from monetary liability for failure to perform a contractual obligation* where the failure to perform is due to circumstances beyond his control. The Support Agreement involves a totally different situation: the *"quid pro quo"* of support for information. Thus, if the Supporting Party has received little or nothing of value under the agreement, he should not be required to pay the support.[11]

Under today's complicated drilling conditions, consideration should be given to providing for up to a *total earning* if all but a few "technicalities" of the earning conditions have been complied with, and for a *partial earning* if information of significant value concerning interim formations has been provided the Supporting

[11] Note that general Force Majeure provisions would not be appropriate in Support Agreements, since most of the "agreements" on the part of the Drilling Party are normally not contractual obligations (*covenants*), but merely *conditions of earning* the support. See § **2.06(a)**.

Party, even though the earning well could not be drilled to total depth, or certain tests could not be performed—*so long as* the failure literally to comply is due to circumstances beyond the control of the Drilling Party.

The drafting of such provisions will not be easy, and must be hand-tailored to each particular deal. For instance, what constitutes "substantial" compliance? Should the support be prorated, and if so, on what basis? What conditions are in fact "beyond the control" of the Drilling Party? For instance, what if the failure to literally comply is correctable, but only at an excessive cost? Is the fact that a well is running far over original cost estimates grounds for not continuing the drilling? What if an error results not totally without fault on the part of the Drilling Party, but is the result of "simple" negligence on the part of one of his employees?

For a fairly inexpensive well, and fairly minor support, the time and effort required in thinking through, negotiating, and drafting such a complicated provision is probably not worth it. Where hundreds of thousands or millions of dollars are at stake, a more careful consideration of the problems clearly seems called for.

§ 2.04(d) Bottom Hole/Dry Hole Provisions. As mentioned in **§ 2.01(c)**, Support Arrangements may either be *"Bottom Hole"* or *"Dry Hole"* in nature. While Bottom Hole Contributions are far more valuable from the standpoint of the Drilling Party, they are also less common.

Where the Support Agreement provides for a Dry Hole Contribution, serious problems arise in determining what is a "dry hole," or nonproductive well. If the Drilling Party plugs a well in good faith, but the Supporting Party feels that the well was not a dry hole, the support may be lost. On the other hand, a Drilling Party has often been known to plug a well which could have been completed as "commercial" where the value of the support money to be earned under a Dry Hole Contribution exceeded the estimated present worth of the future net revenues from the well.[12]

If a Dry Hole Contribution is being made toward a well where shallower, previously-developed "mature" formations exist in the area which are likely to be productive, the Drilling Party must insure that the condition of earning is that the well not be productive *in depths below* the base of the likely-to-produce formations.[13]

It is important that the standards for determining whether or not a well meets the Dry Hole requirements be clearly defined in the contract. The test should not be whether the well is plugged and abandoned, but whether or not the well establishes "commercial production," which should also be clearly defined.[14]

One test would be to provide that the support will be earned depending on whether or not the well is "commercially productive in accordance with customary practices in the industry." While this would be a reasonable test, it is highly subjective, and one on which the parties again might find themselves in serious disagreement—good faith or otherwise. (Compare § § **1.01(d)** and **2.04(b)[2]**.)

From the standpoint of the Supporting Party, the best provision would be to provide that the support would be earned if the well is plugged as a dry hole "with the approval of the Supporting Party." While this certainly provides maximum protection to the Supporting Party, it would also permit him to "stiff" the Drilling Party. Thus, from the standpoint of the Drilling Party, any such provision should be modified by adding "which consent shall not be unreasonably withheld." Again, this modification results in a highly subjective standard.[15]

---

[12] Problems also arise, for Dry Hole purposes, in determining where "testing" ceases and "production" begins.

[13] As to other special requirements for the support of a test to a new horizon, see § **2.03(b)[3]**.

[14] As to such definitions, see **Chapter Eight.**

[15] The Support Agreement might provide, if necessary, for resolving any dispute through tests by an independent testing firm, or through arbitration.

Typically, as in other "earning condition" areas (see § **2.04(c)**), Support Agreements have been on a black/white basis in establishing the dividing line between Dry Hole and Bottom Hole Contributions. However, frequently a major concern for the Drilling Party is not merely the risk that the well will be *non-productive*, but that the well will not prove *profitable*. Many wells which are "commercial producers"—i.e., justify completion—will never payout or prove profitable. (See § **1.03(c)[2][a]** concerning " 'Yet to Spend' Economics.") Serious consideration should be given to a "mid-ground" support which would provide for payment of all, or at least a portion, of the contribution if a *marginal well* were completed, or if the well did not meet certain standards concerning minimum monthly production or attained certain results on testing. However, from the standpoint of both parties, such support should be payable (or refundable if previously paid) if the minimum requirements were not met over the specified period of time, e.g., three to six months.

If a Bottom Hole Contribution is given, the Supporting Party should also consider adding a provision under which such support could be recouped out of a percentage of future production from the well. While such an arrangement reduces the value of the support to the Drilling Party, it provides him with the non-recourse "up front" money for the drilling of the well and permits him to "repay" the support over a period of time if the well is productive.

As to the tax consequences of such a recoupment provision, see § **2.07**.

As to "incentive" arrangements where Dry Hole Contributions convert to Bottom Hole Support in the event of certain performance by the Drilling Party, see § **2.03(b)[4]**.

§ 2.04(e) Insuring Compliance with the Earning Conditions. In many instances, a right to support is inadvertently lost due to an unnecessary failure by the Drilling Party's on-site personnel to comply with the earning conditions. Frequently, the drilling crew will not be aware of various provisions in the Support Agreement requiring notification of the Supporting Party's personnel at various stages of the operation (see § **2.04(b)[3]**), or requiring certain tests that might otherwise not have been performed by the Drilling Party (see § **2.04(b)[2]**). Since *strict compliance* with these provisions will normally be a condition of the support (see § **2.04(c)**), and often is also a part of the spirit of the agreement, an inadvertent loss of the support through such a failure to comply, is, of course, to be avoided. Accordingly, it is critical that the drilling superintendent be fully advised concerning the requirements of, and the necessity for compliance with, the various earning conditions.

§ 2.05 "Take-Over" Provisions. Provisions permitting the Supporting Party to take over the earning well to conduct his own tests are quite common in the Support Agreement. (See § **2.04(b)[4]**.) However, Support Agreements normally do not give the Supporting Party a right to take over the earning well if the Drilling Party elects not to complete—a common provision in Farmout Agreements. (See § **3.04(c)**.)

In the Farmout Agreement, the acreage being "taken-over" by the Farmor is acreage which is yet to be earned by the Farmee. In the Support arrangement the problem is more complicated, since the acreage involved is that of the Drilling Party.

Under these circumstances, it may be that no take-over provision is called for. However, some Support Agreements are now providing a *"Takeover" Right* for the Supporting Party, *so long as* the necessary contribution is made.

In some instances, a Supporting Party has requested a right to take over the well with no obligation to pay the support under a Dry Hole arrangement unless it is unable to successfully complete. While such a position may be understandable, it is not one to which most Drilling Parties would willingly consent.

If the Supporting Party feels that the well could be successfully completed as not merely a "commercial" well but as a profitable one as well, he should not mind paying the support, even in a Dry Hole arrangement, for the right to take over the well. Another alternative would be to permit him to take over the well, assigning an Overriding Royalty Interest to the Drilling Party, or a right to recoup his drilling costs out of production (any right to take over should, as in the case of the Farmout Agreement, be limited to the formation in which the well is completed: see § **3.04(c)**.

While the right to take over is meaningful to the Supporting Party in those instances where he and the Drilling Party disagree as to the potential profitability of the earning well, it is not necessarily a solution for the problems discussed in § **2.04(d)** where the support must be paid in a Dry Hole arrangement as a condition of taking over the well, since the cost of the support may be far more than the value of the reserves to be realized even if the well is completed as "commercial." Accordingly, unless the duty to pay any Dry Hole Support is limited to situations where the Supporting Party is unable to successfully complete the well, a proper definition of *Dry Hole* should still be inserted in the Support Agreement (see § **2.04(d)**).

In any event, if the Supporting Party is given a right to take over the earning well, he should be required, as in the Farmout Agreement, to reimburse the Drilling Party for the estimated salvage value of the Drilling Party's equipment.

§ 2.06 Miscellaneous Considerations

§ 2.06(a) Covenant or Condition?[16] Generally speaking, the drilling of the earning well will not be a *contractual obligation* on the part of the Drilling Party, but only a *condition* of earning the support. (Needless to say, the Drilling Party should carefully review the proposed Support Agreement to insure that this is the case.) However, as mentioned in § **2.04(b)[3]**, the Drilling Party should not be permitted to evade the spirit of the Agreement by going "tight hole" on the Supporting Party once he realizes that the earning conditions cannot be satisfied—such as in a Dry Hole Contribution

[16] As to the general distinction between convenants and conditions, see § **1.01(d)**.

where a productive interim formation is encountered—or where it is otherwise in the Drilling Party's apparent best interests not to provide information to the Supporting Party even though it results in a forfeiture of the support.

A Supporting Party has earned his right to the information once he places himself "at risk" by committing to the support, even when that support is Dry Hole in nature. The Drilling Party should not have the election to violate the spirit of the Agreement by determining, in his discretion, that he no longer chooses to provide information. Yet this is just how the typical Support Agreement reads. Accordingly, the Support Agreement should *contractually obligate* the Drilling Party to provide information as to tests actually run; and such provisions should be a covenant, and not a mere condition, of the Support.

§ 2.06(b) Indemnification Provisions. With the limited amount of "control" normally granted a Supporting Party under a Support arrangement it is highly unlikely that the Supporting Party would be deemed a "mining partner" and thus potentially liable for any damages to third parties resulting from the drilling of the earning well. Nevertheless, the Support Agreement will provide that the Drilling Party will drill the earning well at his sole cost, risk, and expense; and he will indemnify the Supporting Party against any liability.

Indemnification provisions also need to be included in the Support Agreement under which the Supporting Party will indemnify the Drilling Party where the Supporting Party has taken over the earning well to conduct his own tests or to make a completion attempt. (See § § **2.04(b)[4]** and **2.05**.) Frequently, such provisions are inadequately drawn, and merely indemnify the Drilling Party against *the costs of such operation.* These provisions do not adequately cover possible exposure to *liability to third parties* for damages resulting from the conduct of operations by the Supporting Party when he is in control of the well, or to *a loss of the hole.*

At a minimum, the indemnification provisions should cover all three potential areas of loss to the Drilling Party:

(1) The expenses of the "take-over" operation;

(2) Liability to third parties;

(3) Loss of the hole.[17]

The Drilling Party may also wish to consider a provision under which he has a right to refuse to permit the Supporting Party to take over the well under specified circumstances, e.g., where the conduct of certain tests, such as velocity surveys, or a completion attempt in a lower zone, might lead to a loss of previously established reserves in an upper horizon.

§ 2.06(c) Confidentiality of Information. If the information concerning the drilling of a well is valuable enough to warrant support, serious consideration should also be given to inserting a "confidentiality of information" provision in the Support Agreement.

Serious questions exist as to the effectiveness of such provisions; how proof can be made of their violation; and what an appropriate measure of damages would be if the provision is violated and this violation can be proved. However, inserting a confidentiality of information provision in the Support Agreement, particularly where this may serve to keep the information confidential *at least for a limited period of time,* in most instances will prove beneficial to both the Supporting and Drilling Parties.

§ 2.06(d) Area of Mutual Interest Provisions. Traditionally, Support Agreements do not contain Area of Mutual Interest ("AMI") provisions. However, the same reasons that lead the parties to a Farmout or Joint Operating Agreement or Seismic Option, to insert such a provision in their arrangement may under certain circum-

[17] Any loss-of-hole indemnification should cover only the cost of drilling a new well to the same depth, and not the loss of any speculative value from lost reserves, including reserves already established in upper formations.

stances make an "AMI" provision an intelligent addition to a Support Agreement. Accordingly, where there are large amounts of open acreage in the general area of the Supporting and Drilling Parties' leases, consideration should be given to the possibility of adding an Area of Mutual Interest arrangement.

Area of Mutual Interest provisions are discussed in detail in **Chapter Seven.**

§ 2.06(e) Other Provisions. In some instances, a Supporting Party has made a Bottom Hole or Dry Hole Contribution toward a well, anticipating a spacing pattern that would limit the Drilling and Spacing Unit for the earning well to the acreage owned by the Drilling Party. After the well is brought in—if the well is mediocre, or in some instances, while the well is drilling—the Drilling Party may apply for and secure a larger spacing which brings in the Supporting Party's adjacent acreage; and then may force pool the Supporting Party into the earning well (and his costs).

To alleviate the adverse effects of such an occurrence, the Support Agreement may provide for consultation and/or approval of the Supporting Party toward any spacing application filed by the Drilling Party. While the Supporting Party cannot prohibit a greater spacing on the conservation agency's motion, or upon the application of a third party, under certain circumstances the contributions still could be taken into account in determining the Supporting Party's "forced" cost share, or the recoupment of any "penalty" against his interest.

In some circumstances, a Supporting Party has been successful in insisting on adding a "call on production" to the Support Agreement, either as a condition of his support or as a firm contractual obligation. Whether or not the Drilling Party will be willing to accept such a proposal is simply a matter of negotiation between the parties and the value which the Drilling Party places on the support.

§ 2.07 Tax Consequences of Support Contributions. The position of the industry has always been that a Support Contribution reduces the Drilling Party's Intangible Drilling and Development Costs, which has the effect of reducing any Productive-Well Intangible Drilling and Development Costs incurred in connection with such a well.[18] Arguably, however, the Support Contribution should be prorated between IDC and Equipment Costs. (Since IDC deductions are more valuable than ACRS recovery and Investment Tax Credits on equipment, this would be less beneficial to the Drilling Party.)[19]

Insofar as the Supporting Party is concerned, the industry has taken the position that, while Bottom Hole Contributions *on productive wells* must be capitalized as a Lease Cost (in the nature of Geological and Geophysical Expenditures) of the benefited leases. Dry Hole Contributions, and Bottom Hole Contributions on dry holes are currently deductible.[20]

It is now the position of the Internal Revenue Service, as reflected in Revenue Ruling 80-153, that a Bottom Hole Contribution is "income" to the Drilling Party. (Since such income would be non-depletable, this would be less advantageous to producers entitled to Percentage Depletion due to the "50% net income limitation." See § **1.03(b)[4]**. The IRS in this Revenue Ruling also took the position that a Supporting Party must capitalize Bottom Hole Contributions even where the earning well is a dry hole, unless the Supporting Party also releases his surrounding leases.[21]

The question remains as to whether the IRS will soon formally take the same position on Dry Hole Contributions as its "Bottom

---

[18] As to IDC deductions and Productive-Well IDCs, see § § **1.03(b)[6]** and **1.03(b)[7]**.

[19] As to the Investment Tax Credit and the ACRS method of recovering equipment expenditures, see § **1.03(b)[5]**.

[20] As to the capitalization of Geological and Geophysical Expenditures as Lease Costs, see § **1.03(b)[4]**.

[21] As a Lease Cost, such capitalized expenditures would be the *least valuable* to the Supporting Party for Federal income tax purposes.

Hole" position reflected in Revenue Ruling 80-153.

Where a Supporting Party provides for a recoupment of a Bottom Hole Contribution out of production (see § **2.04(d)**), such recoupment will probably result in the contribution being classified for Federal income tax purposes as a *Production Payment.*

A Production Payment is an interest in minerals in place that entitles the owner to a specified fraction of production, which is limited to a specified time or amount (either in dollars or units of production). The tax treatment of such Production Payments varies, depending on whether or not the Production Payment is *pledged to the development* of the property.

Bottom Hole Contributions recoverable out of production would be treated as Production Payments pledged to development; if so, as a result:

(1) *From the standpoint of the Drilling Party,* he would realize no income as a result of the contribution, but must reduce his Intangible Drilling and Development Costs and equipment expenditures by the amount of the contribution. The Drilling Party will *not* be taxed, however, on the revenues that are utilized to repay the contribution.
(2) *From the standpoint of the Supporting Party,* he will receive no deduction if the well is productive, though an Abandonment Loss will be allowed if the well is dry. Likewise, the Supporting Party will receive *taxable income* subject to Depletion as his contribution is repaid.

The economic impact of providing that a Bottom Hole Contribution is recoverable out of production is that it obviously would be less favorable to the Drilling Party than if no recoupment were provided for. (See § **2.04(d)**.) However, the Drilling Party, if he qualifies as an "Independent Producer," will at least be entitled to receive increased Percentage Depletion in the year in which the well is drilled over the treatment that would be provided under Revenue

Ruling 80-153.[22]

From the standpoint of the Supporting Party, the right to recoup the contribution out of production obviously means that, economically, he will be better off if the earning well is productive. However, if the earning well is dry, he will receive more favorable tax treatment than would be provided under Revenue Ruling 80-153, since he will be given an Abandonment Loss equal to his basis in his Production Payment without being required to release his leases.

[22] As to the availability of Percentage Depletion to an Independent Producer, see § **1.03(b)[4]**.

Chapter Three

FARMOUT AGREEMENTS

§ 3.01 Basic Concepts

§ 3.01(a) In General. You own a significant acreage position in a prospect area. But, based upon your economic analysis of the acreage, it is not intelligent for you to currently undertake the cost and risks of drilling on this acreage. You may have determined, based upon the "mature" nature of the area, that the anticipated return does not meet your normal return objectives. Or, it can be that the acreage, if productive, promises significant rewards; but, *after adjustment for risk*, the prospective return is substandard. However, if you could transfer the risks of *initial* drilling to another party, who would thus bear the higher risks of testing the area, the costs versus the risks of *further developing* the acreage, if productive, might prove extremely attractive.

The reverse can also be true. You have a prospect idea which your exploration department considers quite worthwhile. The leases in the area, however, are controlled by someone else. Equally possible, you have significant acreage in the area; but if you were to drill on your acreage, you would prove up someone else's acreage, or lose the opportunity to acquire an interest in that acreage from someone who is currently unwilling or unable to bear the costs and risks of initial drilling.

What we are looking at are two sides of a single coin. Just as in the stock market, for each *seller* of the stock who has determined that the time and price are right to sell, there must be a *buyer* who feels, that at the current price, the stock is an attractive buy. Similarly, here the answer for both parties may be a *Farmout Agreement* — and, in any Farmout arrangement, for every Farmor, there must be a Farmee.

The term "Farmout" is generic, and covers both relationships in which a party is assigning, or accepting the assignment, of an interest in the acreage. The term "Farmin," however, specifically refers to a relationship where a Farmout is being *accepted* by a Drilling Party.

There are numerous reasons why I may be willing to farmout acreage, while you are willing to farmin. My return standards may differ from yours; or you and I may disagree as to the risks and/or potential return from the drilling. We may further be in different conditions concerning availability of rigs, lease expiration dates, and budgetary availability or constraints. Whatever the reason, one of us is willing to drill, while the other is not.

Prior to the Economic Recovery Tax Act of 1981 ("ERTA"), the different positions, tax-wise, of a major oil company and an Independent Producer frequently meant that a project that was uneconomic for the major might make very good economic sense for the Independent. The Independent Producer could claim Percentage Depletion; the major oil company could not.[1] The Independent Producer received one treatment under the Crude Oil Windfall Profit Tax; the major oil company was less favorably treated.[2] And, most important, the Independent Producer was normally drilling, at least in part, with monies provided by so-called "tax-oriented investors," whose dollars were being taxed at a Federal income tax rate of 70%, while the major's income was being taxed at a maximum rate of 46%.[3] Under these circumstances, a project which, after risk

---

[1] See **§ 1.03(b)[4]**.

[2] **Id.**

[3] See **§ 1.03(b)[2]**.

adjustment, might be a marginal for a major, could make tremendous sense for the Independent.

Today, many of the tax discrepancies between major oil company and Independent drilling have been eliminated. While Percentage Depletion still remains available to the Independent, but not to the major, the Windfall Profit Tax will normally not be a distinguishing factor on post-1978 drilling.[4] Likewise, as the result of ERTA, both corporations and individuals are taxed at virtually the same maximum Federal income tax rate. While the major — but not the Independent — must capitalize 15% of his Intangible Drilling and Development Costs, the rapid amortization of these costs makes this major/Independent distinction relatively unimportant to the economics of a deal.[5] Future decisions to farmin or farmout, then, will be based more upon differing opinions concerning the economic potential of a given prospect, differing return standards, or differing abilities to control costs.

Most major oil companies, and many Independents, have a sizable investment in leases, in the form of both bonuses and delay rentals and in seismic and other "prospecting" expenditures. However, the owner of such leases may not be in the position to currently develop them, due to budgetary limitations, a need to spread the costs and risks of drilling, or a belief that such leases possess insufficient economic potential compared with other available drilling opportunities. Another party, though, may feel that such leases offer an excellent economic opportunity.

From the standpoint of the first party, there is a need to enter into an arrangement where some other member of the industry will bear the costs and risks of an initial drilling evaluation of the area. From the standpoint of the second party, there is a need to acquire an interest in the leases.

---

[4] See **§ 1.03(b)[4]**.

[5] See **§ 1.03(b)[6]**.

As indicated in § **1.02(c)**, a Farmout Agreement is a contractual relationship under which the party with the acreage position agrees to assign an interest in his leases to a party who is willing to perform certain drilling and testing obligations, and who will permit the assigning party to retain an interest in the leases assigned.

The basic objective of the Farmee in entering into a Farmout relationship is to acquire an interest in these leases at the least possible cost to himself.

The objectives of the Farmor may be much more complex. His basic objectives may include one or more of the following:

(1) To *maintain his leases* in effect.

(2) To have the *area evaluated and tested.*

(3) To secure a *cost-free interest in production* by virtue of his reserved interest under the Farmout.

While all of these objectives may be of some importance to the Farmor, *one* of the objectives will usually be of premier significance. *How* the Farmout Agreement as structured will depend upon which of the objectives is the most important to the Farmor. For instance, if *maintaining leases in effect* is of critical significance, retention of deep rights and the scope of drilling obligations may be the most important "deal points" from the Farmor's standpoint. If having the area *evaluated and tested* is the Farmor's key objective, the location of the earning wells and the testing obligations, together with the Contract Depth provisions, will be key points in the negotiation. On the other hand, if the Farmor believes the area to be of limited interest, and is essentially attempting to salvage as much of his prior investment as possible, he will primarily negotiate as to the size of his retained cost-free interest and the drilling obligations to be imposed on the Farmee.

It is rare that any party to a negotiation can secure each point that he might wish to attain. It is critical, thus, in negotiating a Farmout, that the Farmor's representative clearly understands which of the foregoing objectives are most critical to management, so that he can know where to "give," and where to "hang tight," in the negotiations.

The basic rationalé behind Farmout and Farmin relationships has been discussed in § **1.02(c)**. Insofar as the decision to farmin or farmout versus other possible arrangements concerning the processing of acreage, the following factors are significant:

(1) A **Farmout** makes sense when:

(a) You have *expiring acreage* and are subject to *budgetary restraints* or *lack of rig availability;* or

(b) You wish to *reduce or eliminate risks* and *improve economics* in the sense of return on investment, and are willing, in return, to accept a *reduced acreage position* in the area.

(2) A **Farmin** will make sense from your standpoint when:

(a) You are under *no budgetary restraints;* and

(b) You want to *increase your acreage position* in an area and are willing to accept the *greater costs and risks* (coupled with *reduced return on investment)* for a greater *aggregate* profit.

(For further discussion of circumstances under which a Farmout or a Farmin may make sense, see § **1.04(a)**.)

This chapter will discuss Farmin and Farmout relationships in a relative vacuum, based upon the assumption that your decision is either to Farmout your entire interest in an area, or to accept a total Farmin relationship. Under certain circumstances, you will wish to deal with acreage over a broader area, and will couple a Farmin or Farmout with the formation of a Working Interest Area (see **Chapter Four**); or you may wish to grant or accept a Seismic Option (see **Chapter Five**). In any event, if your Farmout relationship involves more than a single Drilling and Spacing Unit, you will normally wish to couple the Farmout with a Joint Operating Agreement (see **Chapter Four**), which may or may not include an Area of Mutual Interest provision (see **Chapter Seven**.) In any event, there will be a need to define in your Farmout Agreement many contract terms which are frequently used in Farmout Agreements without any clear specification of the exact meaning of the contract terminology (see **Chapter Eight**).

§ 3.01(b) Areas of Negotiation in a Farmout Arrangement

§ 3.01(b)[1] In General. In negotiating a Farmout arrangement, the parties will primarily be concerned with three basic areas of negotiation. First, the extent of the *acreage to be committed* to the deal by the Farmor, assuming maximum performance by the Farmee (the "subject matter" of the Farmout); second, *what the Farmee must do* to earn his interest (the "earning conditions"); and, third, what *interest will be assigned* to the Farmee and what *interest*, and other rights, will be *reserved* by the Farmor ("interests assigned and reserved"). In addition to these three basic considerations, there are numerous miscellaneous and implementing considerations which must be taken into account in negotiating the agreement.

While all of these points, plus numerous contractual "details,"[6] will eventually be covered in the formal Farmout Agreement, there are certain points which are key to the negotiation of the Farmout and must be agreed upon before it can be determined that any real "meeting of the minds" has occurred.

§ 3.01(b)[2] Subject Matter of the Farmout. The Farmor must first determine the *maximum commitment* that he is willing to make to the deal. In many instances, the Farmor will have considerable acreage under lease in an area, all or only a part of which may be committed to the Farmout. The Farmor must then determine the *maximum amount of acreage* which he is willing to commit to the deal, assuming that the Farmor is willing to perform maximum drilling (see **§ 3.01(b)[3]**). For example, if the Farmor has leases covering 80% of a 12,000 acre area, he could commit his full acreage

---

[6] In certain instances some of these "details," such as whether or not the deal will be structured as a *Tax Partnership* (see § **3.04(b)[2][c-5-b]** *et seq.*), can be enough to make or break a deal, but unfortunately will not have been covered in the negotiations.

block to the Farmout and, accordingly, negotiate a far greater drilling commitment on the part of the Farmee. However, if the Farmor is himself relatively aggressive insofar as drilling is concerned, he might prefer to commit a minimum amount of acreage — say, 3,000 acres — to the Farmout. While it would not be possible to negotiate anywhere near the drilling obligations on the part of the Farmee to earn an interest in this limited amount of acreage, it would leave the Farmor free to conduct a greater amount of "own account" drilling on his retained acreage if an initial test well by the Farmee established production; or it might permit the Farmor to negotiate even more advantageous Farmouts with other parties on the balance of the acreage.

Irrespective of the amount of acreage which the Farmor is willing to commit to the deal, he must also make a decision concerning the *maximum depth* to which such acreage will be committed, again assuming maximum performance on the part of the Farmee. As discussed in § **3.01(b)[3]**, the depths to which the committed acreage will be assigned will normally be dependent upon the depths drilled by the Farmee in his earning wells. However, irrespective of the willingness of the Farmee to drill to greater depths, there may be certain deep formations which the Farmor wishes to retain — depths which he does not wish the Farmee to earn even if the Farmee is willing to test such deep horizons. The Farmor must accordingly balance his wishes to see such depths tested with his counter desires to retain such deep rights for subsequent evaluation and exploitation through his own efforts and for his own account.

While decisions concerning the maximum amount of acreage to be committed to a deal (the *Farmout Area* decision), and the maximum depths to be committed (the *Farmout Depth limitations*), are critical to the commitment decision, there are other questions in defining the subject matter of the agreement that are also subject to negotiation, and to which the parties should have given thought prior to entering into the negotiations. For instance, is the Farmor willing to make any Bottom or Dry Hole Contribution, or share pre-existing seismic or similar information as a "sweetener" to encourage the Farmee to accept a high-risk Farmout? Is the Farmor going to require any *cash consideration* for the assignment of his leases in addition to the drilling and testing requirements? And, finally, what agreement, if any, are to be made concerning any *Subsequently-Acquired Leases,* including both leases acquired in or near the Farmout area by the Farmor or by the Farmee?

The Subsequently-Acquired Lease decision may vary according to the circumstances under which the leases are acquired. The Farmee will frequently take the position that if leases are acquired within the Farmout Area by the Farmor during the "earning period" — the period during which the Farmee's earning wells are being drilled — then the Farmee should be entitled to an interest in such leases irrespective of whether or not the parties want to include a general *Area of Mutual Interest* provision in the Farmout Agreement. In some cases, the Farmee will be willing to pay his proportionate part of the cost of acquiring such leases; in other instances, he will argue that the "acquisition costs" of such leases are, in effect, being paid for by the drilling of the earning wells.

Regardless of the decision concerning leases acquired by the Farmor during the earning period, and even though the parties may have determined that neither of the parties wants their agreement to be subject to general Area of Mutual Interest provisions, if any of the leases committed to the Farmout Agreement are subsequently extended or renewed, the parties may wish some type of limited "AMI" rights to apply to such *extensions and renewals* of the Farmout Leases.

Finally, a decision must be made as to whether a more general Area of Mutual Interest provision should be included in the Farmout Arrangement (see **Chapter Seven**).

§ 3.01(b)[3] The Earning Requirements. Irrespective of the maximum amount of acreage and the maximum depth commitment which the Farmor is willing to make to the Farmout Agreement (see

§ 3.01(b)[2]), the Farmee's right to earn such interests will be dependent upon the *extent of his performance* under the Farmout Agreement.

The Farmee's earning requirements will establish both a *minimum standard* which he must satisfy if he is to earn *any interest* under the Farmout Agreement, and numerous *additional "options"* that must be satisfied if he is to earn his *maximum interest* under the Farmout. A first consideration is the *wells to be drilled* for the Farmee to earn. *How many* wells must be drilled, either as a condition of maximum earning, or as a minimum earning requirement? Are such wells to be "optional" (mere *conditions* of the earning), or are they to be "mandatory" ("required" or "mandatory wells"), i.e., wells whose drilling are *covenants* (firm contractual obligations) on the part of the Farmee?[7] As to each well, how much acreage will be earned by each?[8]

Under the typical exploratory Farmout, it is not merely the drilling of a specified number of wells that is necessary to fulfill the earning requirements; the *depths* to which such earning wells are drilled will also affect both the fact of earning and the depths to which the Farmee will earn.

Normally, in the exploratory Farmout, the Farmor will wish to establish a *minimum Contract Depth* to which each earning well must be drilled if the Farmee is to earn anything. From the Farmee's standpoint, he will also insist that the Farmout Agreement contain certain *provisions excusing further drilling* under circumstances where the Contract Depth cannot be attained. Both parties will want the agreement to specify whether such excuses result in no, total, or a partial earning, either with or without some type of "substitute well"

---

[7] As to the distinction between covenants and conditions, see § **1.01(d)**.

[8] As to "sweeteners" which may be added to the agreement if the Farmee in fact drills the *maximum* number of earning wells, see § **3.03(b)[6][c]**.

provision.[9]

The Contract Depth provisions merely specify the minimum depths to which the Farmee must drill if he is to earn anything. In addition, the Farmout Agreement must specify the *Farmout Depth* — the depth to which the Farmee will actually earn — so long as the earning wells are drilled to the minimum *Contract Depth.*[10]

In addition to the foregoing requirements, the Farmor may wish to impose certain additional earning requirements concerning the wells to be drilled. These may include a "productive to earn" requirement or a "continuous drilling" obligation.[11]

In the exploratory Farmout, particularly where a chief objective of the Farmor is the evaluation of the Farmout and surrounding areas (see § **3.01(b)[1]**), *testing requirements* may be as significant as the requirements concerning the earning wells. Such testing requirements, generally speaking, will follow the same guidelines discussed in connection with the Support Agreement. (See § **2.04(b)** *et seq.*) However, in the Farmout Agreement, the Farmor may frequently be able to impose more substantial testing obligations, and greater rights for conducting his own tests, than a Drilling Party might be willing to accept in a Support arrangement.

In a "rank wildcat" Farmout, the lack of any area "control" may be a significant motivation for the Farmor's decision not to drill for his own account. Accordingly, in such Farmouts the Farmor may wish to impose certain *seismic obligations* as a preliminary and additional requirement of the earning.

§ 3.01(b)[4] Interests Assigned and Reserved. The amount of acreage earned by the drilling of each earning well is a part of the

---

9 As to Contract Depth and related Farmout Agreement provisions, see § **3.03(b)[4]** *et seq.*

10 As to Farmout Depth provisions, see § **3.03(b)[5]** *et seq.* As to limitations which the Farmor may wish to place on the Farmee's ability to earn, see § § **3.01(b)[2]** and **3.03(b)[5][d]**.

11 As to such additional earning well requirements, see § § **3.03(b)[6][b]** and **3.03(b)[7]**.

"earning requirement" considerations discussed in § **3.01(b)[3]**. However, the *interest in acreage and production* to be assigned to the Farmee, and the similar interest to be retained by the Farmor, are other key items of Farmout Agreement negotiation.

Conventionally, the Farmor will assign to the Farmee 100% of the Working Interest (the cost-bearing interest) in each earning-well Drilling and Spacing Unit, reserving an Overriding Royalty Interest, until Payout (the point at which the Farmee has recovered all of his costs of drilling, completing, and equipping the earning well, plus operating it to the "payout" point); thereafter, the Farmor's Override would be convertible, at the Farmor's option, to a Working Interest.[12] As to the "outside excess" acreage — the additional undrilled Drilling and Spacing Units earned by the Farmee by the drilling of each earning well — the Farmor would assign an undivided Working Interest to the Farmee in such acreage, also retaining an undivided Working Interest in the acreage.

Even in the conventional Farmout Arrangement, the *size* of:

(1) The Farmor's Overriding Royalty Interest in the earning-well Drilling and Spacing Unit;

(2) The convertible Working Interest; and

(3) The Working Interest to be assigned and reserved in the excess acreage;

were subjects of negotiation which would vary from area to area, and would also vary depending upon the costs and risks associated with the drilling of the earning wells. Furthermore, the Farmor and Farmee were also free to negotiate "non-conventional" arrangements such as disproportionate cost-sharing or casing-point elections. And in recent years, the occurrence of such non-traditional cost-sharing relationships has proliferated.[13]

---

[12] As to the tax reasons for this "conventional" Farmout arrangement, see § **1.03(b)[6]**.

[13] As to the reasons for such "new" arrangements, and the alternatives available to the parties, see § § **3.04(b)[2][c-4-b]**, **3.04(b)[2][c-4-c]**, **3.04(b)[2][c-5-b-4]**, and **3.04(b)[3]** *et seq.*

Recent tax developments, and the desire for more innovative structuring arrangements, have also made the question of whether or not the Farmout would be structured on a *Tax Partnership* basis a significant matter for Farmout negotiation.[14]

Traditionally, the interests assigned and reserved in the "excess" acreage outside the earning-well Drilling and Spacing Units would be *undivided* Working Interests. However, in certain circumstances, the Farmor might wish to retain an Overriding Royalty Interest as to the entire Farmout Area, including the "excess" acreage, or impose a "continuous drilling" obligation if the Farmor was to earn such additional Drilling and Spacing Units.[15] Likewise, under certain circumstances, it may be to the advantage of both parties for the assigned and reserved interests to be alternate 100% Working Interests in a "checkerboard" pattern rather than *undivided* interests throughout such excess acreage.[16]

In addition to the interests assigned and reserved in production and acreage, the Farmor may wish to make other "operational" reservations. These may include "Take-Over" Rights if the Farmee elects not to complete an earning well or to drill to Contract Depth (or the Farmor wishes to conduct his own tests), a right of reassignment if the Farmee elects to "drop" a portion of the assigned acreage,[17] a preferential right to purchase, a call on production,[18] or gas processing rights. (See § **3.04(c)**.)

---

14 As to the advantages and disadvantages of a Tax Partnership, see § **3.04(b)[c-5-b]** *et seq.*

15 As to the continuous drilling obligations, see § § **3.01(b)[3]** and **3.03(b)[6][b]**.

16 As to the reasons for "checkerboard" assignment, see § **3.04(b)[3][c]**.

17 The Farmee may also wish a similar right of reassignment.

18 Under certain circumstances, the Farmee may instead negotiate a "call" on the *Farmor's* share of production.

§ 3.01(b)[5] Implementing Provisions

§ 3.01(b)[5][a] Provisions Concerning Lease Assignment. Traditionally, the conventional Farmout has called for the Farmee to be assigned his interest in the earned acreage only *after* the earning wells have been drilled. But more recently, Farmout Agreements, for tax reasons, have tended to provide for a "present assignment," under which the Farmee is assigned or "subleased" his interest in the acreage to be earned *at the time the Farmout Agreement is entered into*, subject to a duty to reassign (or an automatic reassignment or sublease termination) if the earning conditions are not satisfied.[19]

Regardless of the timing of the assignment, the parties must also determine whether a copy of the formal assignment to be utilized will be attached as an exhibit to the Farmout Agreement, or whether a mere "general" assignment obligation/provision will be included in the Farmout Agreement. (See § **3.05(c)**.)

§ 3.01(b)[5][b] Provisions Concerning Subsequent Joint Operations. Unless the Farmout Agreement is limited to a single Drilling and Spacing Unit, or involves a "checkerboard" arrangement as to any excess acreage, an additional agreement must be reached between the parties concerning how additional operations on the now jointly-owned excess acreage will be conducted.[20] A decision must be reached as to whether a verbatim copy of such *Joint Operating Agreement* will be attached to the agreement, or whether the agreement will be limited to a reference to a Joint Operating Agreement subsequently to be entered into by the parties.

A lawyer will be much happier if the verbatim Joint Operating Agreement is attached as an exhibit to the Farmout Agreement.

---

[19] As to the pros and cons of "present" versus "when earned" assignments, see § **3.04(b)[2][c-5]**.

[20] Even in the single Drilling and Spacing Unit arrangement, an Operating Agreement is in order to cover deepening, plugging back, or reworking operations.

However, time considerations may make this impractical.[21]

While mere reference to a form of Operating Agreement may expedite the execution of a Farmout Agreement, such a practice frequently leads to subsequent disagreement as to exactly *what type* of joint operating relationship is to exist between the parties. For instance, what *form* of Operating Agreement is to be used? While it is normally understood, and frequently expressed, that the Operating Agreement will be on the "AAPL Model Form," what *version* of the AAPL Form 610 is to be used?[22] How are the *blanks and options* under the agreement to be completed? What *deletions* or *modifications* are to be made in the agreement, e.g., deletion of the preferential right to purchase provision? What *accounting procedure* is to be attached?[23]

If it is possible that "split stream connections" may exist in the Farmout Area, an additional consideration is whether or not the parties should pre-agree as to the form of any *Gas Balancing Agreement* to cover such gas deliveries. In many instances, it is better to leave the negotiation of Gas Balancing arrangements until the earning wells have been drilled and splitstream connections are imminent. However, the 1977 and 1982 versions of the Model Form Operating Agreement specifically envision the possibility of split-stream problems, and their possible resolution through an attached Gas Balancing (Storage) Agreement; and the 1982 version even

---

[21] In many companies, "formal" agreements require the approval of a higher level of management, or of multiple parties, which may be impossible to secure prior to the time the earning wells must be commenced. Under such circumstances, practical considerations may require entering into a "simple" letter agreement arrangement, which merely refers to a future Joint Operating Agreement.

As to the enforceability of such "less formal" contracts, see § **1.01(c)**.

[22] Some companies are still entering into Operating Agreements on the AAPL Model Form (1956 version) rather than the more generally used 1977 version. Even greater confusion will arise now that the 1982 version has been adopted by the AAPL.

[23] As to the accounting procedure, it is normally presumed by the parties that a "COPAS" form will be used. But what date? And, how are the blanks and options to be completed?

provides alternate versions of its "page 8," depending on whether or not a Gas Balancing Agreement exists between the parties.

As to problems concerning joint operating relationships, see **Chapter Four**. As to Gas Balancing provisions, see **Chapter Six**.

§ 3.01(b)[6] Other Considerations. A number of other points need to be considered by the Farmor and the Farmee in Farmout Agreements, particularly when considerable exploratory acreage is involved. Do leases need to be extended? Would a *Continuous Drilling* or *Productive to Earn* provision speed up the development of existing acreage prior to lease expiration dates? Should a *fixed drilling schedule* on *unusually tough "Non-Consent" penalties* be adopted for the same reason? Would a *simultaneous drilling* provision as a condition of earning be a "deal point" in view of lease expiration dates? Would *simultaneous Farmouts* to several Farmees make more sense than committing the maximum possible acreage to a single Farmout?

As the deal is being presented to a potential Farmee, what is the risk to the Farmor that the Farmee will use the negotiation period as an opportunity for top leasing the Farmor's leases, or, upon seeing the Farmor's substantiating data, elect to engage in "own account" drilling on the Farmee's own acreage rather than farming out or participating in a Working Interest Area with the Farmor? Do these dangers require the negotiation of a preliminary "anti-top-leasing" or "Farmout or Join" agreement? If mandatory wells are included in the Farmout Arrangement, should a non-drilling payment be included?

Insofar as miscellaneous implementing provisions are concerned, consideration should be given to what *title information* will be furnished the Farmee, and whether the Farmout Agreement will contain any *warranties of titles.* (While General Warranties are unusual in Farmout Agreements, in some instances a Farmor may be willing to agree to a "Special Warranty" that *he has done nothing* to adversely affect title to the Farmout Leases.) Who will pay *Delay*

Rentals? What proof of Delay Rental payments will be provided? How are the *costs* of Delay Rental payments (as contrasted with the administrative duty to pay) to be borne? Will these relationships change after the earning conditions have been performed? Has the Farmee allowed himself sufficient time to complete title work, and to secure the necessary rigs, before the initial drilling must be commenced? Should the parties consider the formation of a Federal Exploratory Unit as a complement to their Farmout arrangement?

§ 3.01(c) Need for Formalizing the Farmout Arrangment. It would be highly unusual for intelligent Farmout negotiations not to cover such areas as the amounts of acreage and depths to be committed to the Farmout Agreement; the number of wells to be required of the Farmee and the acreage to be earned by each; the depth-earned provisions and any special earning requirements such as Productive to Earn or additional seismic; and the interests to be assigned and reserved. However, in many instances, when the formal Farmout Agreement is tendered, the Farmor and the Farmee may learn, to their consternation, that there has been no meeting of the minds as to certain "deal points." These deal points may include AMI rights, either in general or as to extension or renewal leases; Substitute Well provisions; testing requirements; Tax Partnership provisions; operational reservations; Operating Agreement or Gas Balancing provisions; and warranty or delay rental requirements. This is particularly upsetting when, due to lease expiration or similar problems, an earning well has already been commenced under an informal, incomplete, and inconclusive "letter of understanding."

It is understandable that in most cases not every possible point of contention will have been covered in the preliminary Farmout negotiations. Still, whenever possible, it is advantageous if a copy of the proposed Farmout Agreement can be reviewed by the parties, and hopefully agreed upon and executed before any irrevocable commitments to the Farmout arrangement have been made. A subsequent determination that potential "deal points" have not, in fact, been agreed upon can cause a potential Farmee to expend large amounts of money in preparing for a Farmout which is never consummated; it can cause a Farmor with "tight" lease expiration dates to lose his leases.

In some instances, circumstances may dictate proceeding with the Farmout arrangement before an all-inclusive agreement has been executed by the parties. Commencing drilling under an incomplete agreement may prove preferable, from both parties' standpoint, to the loss of leases which have already been top-leased by a third party. However, both the Farmor and the Farmee must realize that drilling a Farmout on a handshake is risky business which can prove costly to both parties.

§ 3.01(d) Farmout Negotiation Checklist. The following checklist summarizes the areas of negotiation discussed in **§ 3.01(b)** and its subsections:

(1) Farmor's Commitment (Subject Matter of the Agreement):
- (a) Maximum amount of acreage to be committed (Farmout Area)
- (b) Maximum depth to be committed (*Farmout Depth* limitation)
- (c) Treatment of Subsequently-Acquired Leases:
  - [1] Leases acquired by Farmor during earning period
  - [2] Extension and renewal leases
  - [3] General Area of Mutual Interest Provisions
- (d) Bottom Hole/Dry Hole or information contributions
- (e) Cash consideration to the Farmor

(2) Earning Requirements:
- (a) Wells to be drilled:
  - [1] Mandatory wells:
    - [a] Number

[b] Acreage to be earned through mandatory drilling
[2] Optional wells:
[a] Number
[b] Acreage to be earned by each
[3] Incentive earning if maximum number of wells drilled
(b) *Contract Depth* requirements:
[1] In general
[2] Excuses from continued drilling and effect on earning
[3] Substitute Well provisions
(c) Special provisions concerning the earning wells:
[1] Productive to Earn
[2] Continuous Drilling
[3] Other
(d) Depths to be earned (*Farmout Depth* provisions)
(e) Other earning requirements:
[1] Testing requirements
[2] Seismic
[3] Other
(3) Interests Assigned and Reserved:
(a) Interests in production and acreage:
[1] Earning-well Drilling and Spacing Unit
[2] Outside acreage:
[a] In general
[b] Undivided interest or checkerboard
[c] Tax Partnership provisions
(b) Operational reservations:

[1] Take-Over Right at completion or if drilling to Contract Depth excused
[2] Right to conduct own tests
[3] Right of reassignment:
[a] Farmor
[b] Farmee
[4] Preferential right of purchase:
[a] Farmor
[b] Farmee
[5] Call on Production:
[a] Farmor
[b] Farmee
[6] Gas Processing rights:
[a] Farmor
[b] Farmee
[7] Other

(4) Implementing Provisions:
(a) Provisions concerning the assignment:
[1] Attach or reference
[2] Present assignment or assignment when earned
(b) Joint Operating provisions:
[1] Attach or reference
[2] Form
[3] Date
[4] Completion of blanks and exercise of options
[5] Accounting procedure
[a] Form
[b] Date
[c] Completion of blanks
[6] Gas Balancing provisions

(5) Special Provisions and Considerations:
(a) Actions to be taken to extend leases

(b) Provisions concerning titles, Delay Rentals and warranty
(c) Non-drilling penalties
(d) Federal Exploratory Unit provisions
(e) Anti-top-lease, or "Farmout or Join" provisions.

§ 3.02 Pitfalls in "Subject Matter" Provisions

§ 3.02(a) In General. "Subject Matter" considerations involve both carefully thinking through what the Farmor is willing to commit to the Farmout (and what the Farmee is willing to accept) — see **§ 3.01(b)[2]** — and properly reflecting these business decisions in the Farmout Agreement. The determination of what rights to potentially commit to the Farmout involves possible limitations on the *lands, leases, depths* and *substances* to be committed to the Farmout.

Normally, the Farmor will expect, and the Farmee will insist, that all leases owned by the Farmor in the *Farmout Area* be committed to the Farmout. The primary consideration concerning any limitations on leases subject to the deal will involve determinations as to rights of both parties concerning Subsequently-Acquired Leases. This topic will be dealt with in **Chapter Seven**. Likewise, limitations on the depths committed to the Farmout (*Farmout Depth limitations*) will be dealt with in **§ 3.03(b)[5][d]**. Similarly, business considerations concerning limitations on the amount of acreage to be included within the Farmout Area have been discussed in **§ 3.01(b)[2]**. This section and its subsections will be limited to a discussion of other restrictions on the rights to be committed to the Farmout.

§ 3.02(b) Limitations on the Farmout Area and the Leases Committed. Once a business determination has been made concerning the maximum acreage to be committed to the Farmout, this Farmout Area must be clearly described in the Farmout Agreement. If any confusion might arise concerning the exact location of the

Farmout Area, a plat should be attached to the Farmout Agreement, clearly defining the boundaries of the Farmout Area. Likewise, if, as is often the case, the leases committed to the deal are to be limited to those portions of the leases lying within the Farmout Area, the Farmout Agreement should clearly so state.

A decision to limit Farmout Leases to that portion of the committed leases lying within the Farmout Area may create complications in lease administration, particularly the payment of Delay Rentals. The Farmout Agreement should clearly specify by whom Delay Rentals will be paid, how determinations will be reached concerning the "dropping" of leases, etc., not only to avoid subsequent disagreements between the Farmor and the Farmee, but also to insure that no lease is inadvertently lost and no lease convenants violated.[24]

Normally, as indicated in § **3.02(a)**, all leases owned by the Farmor within the Farmout Area will be committed to the Farmout. If the Farmor does intend to reserve any or all rights under a particular lease, this should be clearly spelled out in the Farmout Agreement.

The Farmout Agreement should clearly specify the leases being committed to the Farmout. If, as is frequently the case, the list of Farmout Leases is lengthy, the leases will be described on an "Exhibit" to the Farmout Agreement, rather than being set out in the body of the contract. However, where the Farmout Area is extensive, practical problems arise for the Farmee in determining whether or not the Farmor has in fact listed each lease actually owned by him in the Farmout Area.

---

[24] To avoid unnecessarily complicating lease administration, the parties may decide to commit any lease which is primarily located within the Farmout Area to the Farmout as to all acreage covered by the lease, including those minor portions of the lease acreage which lie outside the Farmout Area.

As to comparable problems concerning leases lying partially within and partially outside an Area of Mutual Interest, see **Chapter Seven**.

As to Delay Rental and lease administration generally under the Farmout, see § **3.05(b)**.

The inclusiveness of the enumeration of the Farmout Leases could, of course, be partially verified by a thorough title examination (or at least a quick "record check") of the entire Farmout Area. However, this would be expensive and time consuming, and would not insure that there were not additional leases owned by the Farmor, held in the name of a nominee, which had been inadvertently (or intentionally) omitted from the list of the Farmout Leases. To avoid any such omission, the Farmee should insist that the Farmout Agreement contain a "catchall" — expanding the committed Farmout Leases to include "and any additional leases owned by the Farmor insofar as they lie within the Farmout Area," — subject to the exclusion of any leases which it was mutually agreed were not to be committed to the deal.[25]

An additional problem arises concerning the treatment of Support Contributions to the Farmee from third parties. While the Farmee will normally be credited with 100% of *cash contributions* toward the drilling of the well, with the contribution serving to reduce the Farmee's "Payout" account, the Farmor may or may not share in *acreage contributions*, depending upon the terms of the deal as negotiated. In some instances, the Farmor is permitted to share in acreage contributions without cost. However, the Farmee may insist that if the Farmor is to share in such contributions, he should make a cash contribution equal to the fair market value of his interest in the leases, frequently based upon some pre-agreed-upon dollar per-acre figure.

Certain of the leases committed to the Farmout may be subject to pre-existing gas contracts or other contracts creating obligations, reserving rights, or limiting the freedom of action of the assignee, or which contain unusual clauses or are subject to special restrictions. While the Farmee, as a part of his title verification, should familiarize himself with any such agreements and restrictions,

---

25 The exhibit of leases should still be included to ease the job of the Farmee's Lease Records section.

it is in the interest of the Farmor, as well as the Farmee, to insure that the Farmee does not inadvertently take a course of action which runs afoul of such provisions. Accordingly, to limit avoidable "surprises," it is a good practice for the Farmor to call any unusual provisions or contractual restrictions to the Farmee's attention.

§ 3.02(c) Substance Restrictions. In some instances, a Farmout may be limited to specified substances; certain *rights of the parties* (such as the size of any retained Overriding Royalty Interest) may also vary, depending upon the substances involved. Such restrictions can apply to Oil (or Gas) *Wells*, Oil (or Gas) *Common Sources of Supply*, or Oil (or Gas) *Rights*. Each of these provisions will create different rights and obligations where a well produces both oil *and* gas, or is an Oil (or Gas) Well in a predominantly Gas (or Oil) *Reservoir*. In each case, the Farmout Agreement must carefully define what is meant by "Oil," "Gas," "Oil Well," "Gas Well," etc.[26]

§ 3.03 Earning Provisions

§ 3.03(a) In General. The Farmout Agreement should spell out the *date* by which each earning well is to be *commenced;* the *location* at which each earning well is to be drilled; the *minimum depth (Contract Depth)* to which such wells must be drilled; and the *date* by which each earning well must be *completed*. The *depths earned (Farmout Depth)* by such drilling should also be specified, together with the *number* of earning wells which must be drilled to earn the *maximum* (or *any*) acreage to be assigned to the Farmee, and whether the drilling of such earning wells is *Mandatory* (a "covenant") or *Optional* (a mere "condition" of the earning).[27] The Farmout Agreement should similarly specify the *testing obligations* and *any*

---

[26] As to such definitions, see **Chapter Eight**.

[27] As to the distinction between covenants and conditions, see § **1.01(d)**.

further requirements of the earning. (See § § **3.01(b)[3]** and **3.03(c)**.)

Many of these same matters were topics for consideration in the negotiation of the Support Agreement. (See § **2.04(a)** *et seq.*) However, a Farmout is a more complicated contractual arrangement than a Support Agreement; and, accordingly, the treatment of these subjects in the Farmout Agreement requires more careful consideration and creates additional business considerations and pitfalls.

In defining the *minimum depth* to which any earning wells must be drilled (see § **3.03(b)[4]** *et seq.*) and the *depths* to which *the Farmee will actually earn* (See § **3.03(b)[5]** *et seq.*), the Farmout Agreement often makes reference to a particular formation. It is critical that this formation be defined clearly, so that there will be *no uncertainty* concerning the exact zone intended.[28]

§ 3.03(b) Provisions Concerning the Drilling Obligation

§ 3.03(b)[1] Commencement and Completion Dates. Problems concerning the proper defining of commencement and completion dates, and some of the business considerations to be taken into account in fixing these dates, were discussed in § **2.04(a)[2]**. However, any intelligent setting of a *Commencement Date* must take into account:

1. The *time required* for the Farmee to make *deals with other parties.*

2. The *time required* for the Farmee to *complete* any necessary *forced pooling proceedings.*

---

[28] A common, and excellent, practice is to identify the formation by reference to the log of a particular well, or by reference to *Common Source of Supply* designation in a particular conservation agency order.

Unfortunately, disagreements can still arise between the geological representatives of the Farmor and the Farmee as to whether an objective sand which has previously pinched out has now been re-encountered by the earning well. See § **3.03(b)[4][b]**.

3. The *time required* for the Farmee to complete his *title work.*

4. *Rig availability.*

In fixing both the Commencement and the Completion Dates, both the Farmor and the Farmee must also consider lease expiration dates for both the earning-well Drilling and Spacing Units and any "outside" acreage which will be jointly developed by the parties after the earning wells have been completed, as well as other acreage owned by the Farmor in the vicinity of the Farmout Area. Here, again, rig availability in the area will be a serious consideration; in some instances, a Farmor's decision to farmout to a given Farmee may be based upon the availability of rigs to that Drilling Party.

Fortunately, unlike the Support Arrangement, the Farmor and Farmee, generally speaking, will have a commonality of interest in rapidly commencing and completing the earning wells to permit a full development of the Farmout Area prior to the loss of any Farmout Leases through lease expiration. However, both the Farmor and Farmee may have additional acreage adjacent (or possibly within) the Farmout Area which is separately (not jointly) owned. If such acreage has early expiration dates, the need of one party (but not the other) to secure earning well information quickly may lead to a conflict concerning the fixing of the earning well Commencement and Completion Dates, particularly if it is the Farmor whose acreage is threatened.[29]

§ 3.03(b)[2] Earning Well Location. From the Farmor's standpoint, his reasons for farming out, e.g., expiring acreage versus desires for evaluation of a broad area, will affect his position concerning the location of the earning well or wells. Geological considerations, lease expiration dates, and surface conditions will also affect well location decisions.

---

[29] A similar problem arises in fixing the location of the earning wells. See § **3.03(b)[2]**.

In most instances, the Farmor will want the earning well to be drilled on *his* acreage. However, in certain situations, expiring leases or geological or topographical considerations may lead to the drilling of the well on acreage other than that owned by the Farmor. Such arrangements are referred to as *Off-Lease Farmouts.*

The Off-Lease Farmout may prove an advantage to both Farmor and Farmee where, due to geological or topographical considerations, or because of expiring leases, it is better for the initial drilling to be conducted on the Farmee's acreage. However, the Off-Lease Farmout normally requires the establishment of a Working Interest Area (see **Chapter Four**) for it to work from a business standpoint. (This may also permit the Farmee to earn acreage from several parties with a single earning well under a combined Farmout-Working Interest Area arrangement.) (See § **1.04(b)**.) However, the Off-Lease Farmout also creates special tax problems discussed at § **3.04(b)[2][b]**.

Just as the commonality of interest between the Farmor and the Farmee eases potential disputes concerning well Commencement and Completion Dates, the same commonality makes it easier to arrive at a mutually-acceptable well location or locations. However, both the Farmor and the Farmee may have other acreage in the area which will not be jointly owned and which one or the other wishes to see evaluated, or which contains expiring leases. Since it is the Farmee's money at stake, he may prefer to drill on a location which is optimal geologically, despite a potential lease loss at another location which is somewhat less favorable geologically. Under these circumstances, a legitimate conflict of interest may arise between the Farmor and the Farmee as to the location of the earning wells.

The earning well location can be defined in various ways. The possibilities include:

1. *Any location* within the Farmout Area *selected by the Farmee* ("At a location of the Farmee's choosing . . .").

2. A location of the *Farmee's* choosing *within specified limits*

or as between certain specified alternate locations.

3. *A location* of the *Farmor's choice* (less common).

4. A *"mutually acceptable" location.*

5. A *location specified in the Farmout Agreement.*

If the Farmout Agreement requires only a single earning well, the location of that well will normally be specified in the Agreement. Likewise, even where multiple earning wells are required, the location of the *first earning* (and possibly *all*) *wells* may be specified in the Farmout. However, where multiple earning wells are required under the Farmout, it may prove difficult to determine where the additional earning wells should be drilled until after the first well has been completed and its results evaluated. Similarly, in arrangements where seismic or other "prospecting" work is required as an initial earning condition, it may be premature to designate where *any* of the earning wells should be located until such prospecting information has been gathered and analyzed.

At first blush, it might appear that, due to the "commonality" just discussed, either a provision for drilling "at a location of Farmee's choosing," or at a "mutually acceptable" location, would create no practical problems. However, possible disagreements arising due to conflicts created by wholly-owned acreage, or lease expiration versus geological and monetary considerations, discussed above, may still lead to disagreements under either of these provisions.

While Farmout Agreements frequently provide for additional earning wells to be drilled at locations of the Farmee's choice, such provisions can create problems for the Farmor. Likewise, in the event of serious disagreements between the parties as to well location, a "mutually acceptable" provision is legally unenforceable.[30]

[30] See § **1.01(d)**.

Accordingly, if the acreage position of the parties in the area, or other factors, indicate a likelihood that conflicts could arise, a better provision would be to provide for drilling at a location proposed by the Farmee and approved by the Farmor, "which approval shall not be unreasonably withheld."

As to other problems in defining well location where multiple earning wells are involved, see § **3.03(b)[6][c]**.

§ 3.03(b)[3] Covenant or Condition? A serious point of negotiation between the Farmor and the Farmee will be whether each earning well will be a mere *condition* of the Farmee's right to earn (an "optional" well), or a *firm contractual obligation* or *covenant* (a "mandatory" or "required" well).

Where a single earning well is involved, it may seem relatively immaterial whether or not the Farmee is contractually obligated to drill the earning well. (Why would the Farmee accept the Farmout if he did not intend to perform?) However, if the Farmor is faced with expiring leases, the distinction between covenant or condition can become extremely important from a practical standpoint.[31] Likewise, the dividing line between a mandatory and an optional well will become extremely important where Contract Depth provisions are concerned (see § **3.03(b)[4]** *et seq.*), particularly if unusual drilling conditions and significant cost overruns are experienced by the Farmee. These problems become particularly significant if the Farmee is operating under serious budgetary constraints (or, in the case of a smaller company, finds himself faced with serious financial problems), or where the subsequent drilling indicates that the geological conditions are less favorable than originally anticipated. The imposition of a multiple earning well well requirement (see § **3.03(b)[6]** *et seq.*) further complicates negotiations over *whether* and *how many* earning wells should be mandatory or optional.

If a single earning well is required, more often than not the

---

[31] As to the general covenant/condition distinction, see § **1.01(d)**.

drilling of the earning well will be a mere condition of the earning unless the Farmor is faced with expiring leases.[32] If multiple earning wells are involved, the most common provision is for the initial earning well to be mandatory, and additional earning wells to be optional. However, there are no magic guidelines; and the mandatory/optional nature of all or a part of the earning wells is simply a matter for negotiation between the parties.

As in any negotiation situation, the Farmor must balance his desires to insure contractually that the earning wells will be drilled with the knowledge that a "victory" in this area of negotiation will probably lead to a loss of certain other negotiation points or a necessity for "sweetening" the deal for the Farmee in other regards.[33] If lease expiration is a serious problem, or if the Farmor has questions concerning the Farmee's *bona fides* or ability to perform, the price tag of requiring mandatory wells may prove well worth it. However, due to the reluctance of most oil companies to sue each other over breach of Farmout Arrangements, particularly in view of the subjective nature of many "excuse" provisions (see **§ 3.03(b)[4][c]**), the Farmor may wish to insert a *"non-drilling payment"* in the agreement to give "teeth" to the mandatory nature of the well.[34]

A non-drilling payment may also make sense due to the nebulous nature of the rules concerning damages payable for breach of an obligation to drill. While the Farmee who violates a contractual drilling obligation will be liable to the Farmor for the cost of a well drilled to the required depth, the *proof* of the amount of such damages may prove difficult in a rank wildcat area. Accordingly, to provide certainty for both parties, a fixed non-drilling payment may prove desirable. While the amount of this penalty will be a subject of negotiation, it will usually be somewhat less than the estimated cost

---

32 If the Farmout Area is extensive and "rank wildcat" in nature, it is hard to imagine a situation where lease expiration as to at least a part of the Farmout acreage will not constitute a potential problem.

33 As to Farmout "sweeteners," see § § **3.03(b)[6][c]** and **3.04(b)[3][d]**.

34 If the Farmor has any questions concerning the financial responsibility of the Farmee, he may require that a bond be posted to guarantee the payment of the non-drilling payment.

of drilling the well, but sufficiently large to give the Farmee pause before blithely deciding to ignore his obligation to drill.[35]

From the standpoint of the Farmee, the imposition of any mandatory drilling requirement must also be coupled with adequate "excuse" provisions if the drilling of the well to the Contract Depth proves impossible, impractical, or (if he can so negotiate it) undesirable. (See § **3.03(b)[4][c]**.

While it is less common, certain Farmout Agreements covering exploratory acreage may provide that the drilling of *more than one* (and *possibly all*) of the earning wells will be mandatory. The advisability of such a provision from the standpoint of the Farmee will be discussed in § **3.03(b)[6][c]**.

Frequently, when parties are negotiating an exploratory Farmout covering a broad, multi-prospect "rank wildcat" area of tens to hundreds of thousands of acres, the Farmor may be reluctant to commit rights covering such an extensive area unless the Farmee agrees to drill a number of "mandatory" wells to evaluate the area. Often, over-optimism on the part of the geological department may lead the Farmee to bite off more than he can chew by agreeing to extensive mandatory drilling obligations. A far better provision, from the standpoint of the Farmee, is to agree to a far more limited "mandatory" obligation (say, not more than three or four wells), with additional earning to be drastically limited unless a certain minimum of "optional" wells are also drilled. Another variation would be to excuse the Farmee from further mandatory drilling if a certain number of an initial group of mandatory wells were not productive, or did not meet certain minimum production standards.

§ 3.03(b)[4] Contract Depth, Excuse and Substitute Well Provisions

§ 3.03(b)[4][a] In General. In the exploratory Farmout, the Farmor

[35] In drafting any non-drilling "penalty" provision, the aversion of the courts to "penalty" provisions should be considered. Accordingly, a more euphemistic terminology—such as "liquidated damages" or "non-drilling payment"—should normally be used in the contract.

will normally have in mind some *minimum depth* to which he wishes the Farmee to drill if the Farmee is to earn anything. On occasion, the Farmor may permit the Farmee to earn so long as he establishes production *at any depth*. Under these circumstances, the defining of a minimum *Contract Depth* will merely establish a depth to which the Farmee must drill to evaluate the area *in the event no production is established at a lesser depth* (see § **3.04(b)[4][e]**, below). But an exploratory Farmout often involves certain upper horizons which are expected to prove productive; the true "exploratory" nature of the Farmout is in deeper, as yet untested, zones.

It would be possible for the Farmor to reserve all rights in such upper zones and to make a "sandwich" Farmout (see § **3.03(b)[5][d]**). However, the Farmor will often "throw in" these upper zones as a "sweetener" to entice the Farmee to accept the higher risks of testing still deeper horizons.[36] Under these circumstances, the Farmor will not wish the Farmee to earn, even if production is established, unless the well has been drilled to the specified minimum Contract Depth.

From the Farmee's standpoint, conditions encountered in the drilling of the well may make it impossible, impractical, or undesirable to continue to drill an earning well to the specified Contract Depth. Under these circumstances, he will wish to be excused from further drilling. If the earning well is a mandatory well (see § **3.03(b)[3]**), the "excuse" provisions will be added to the Agreement to avoid a potential "breach of contract" on the part of the Farmee. If the earning well, however, is an optional well, the Farmee could cease the drilling without incurring any contractual liability; here, the excuse provisions will be added to permit the Farmee *to earn* even though the well has not drilled to the Contract Depth, at least under specified circumstances and to a certain extent.

§ 3.03(b)[4][b] Defining "Contract Depth." The Contract Depth

[36] As to other incentives for drilling, see § § **2.03(b)[4]**, **3.03(b)[6][c]**, and **3.04(b)[3][d]**.

provision will normally require the Farmee to drill to a depth sufficient to test (or "X" feet into) a specified formation, or to a specified subsurface (or subsea) footage, *whichever is lesser.*

The first provision is necessary in the event that the objective formations runs "high." (In this event, to drill deeper would be unnecessarily burdensome, and might cause the Farmee to encounter conditions which could actually endanger the formation or the hole.)

The second provision is necessary in the event that the objective formation is simply not present at the earning well location, or would be encountered at a depth so much greater than that anticipated that it would be unreasonable to expect the Farmee to drill to that depth.

A third alternative — a *dollar,* rather than a *footage,* obligation — is discussed in § **3.03(b)[4][c]**.

In designating the "formation" standard, it is imperative, if at all possible, that the formation identified be an *established geological marker,* so that there will be no disagreement as to whether or not the formation has been reached. See § **3.03(b)[5][b]**.

§ 3.03(b)[4][c] "Excuse" Provisions. In drilling the earning well, the Farmee may encounter conditions such as heaving shale, impenetrable substances, loss of circulation, or mechanical difficulties which make it impossible or unreasonably (or unexpectedly) expensive to continue the drilling. The early drilling may also establish that it is unlikely that the well will prove productive in the deeper zones or that it may be dangerous to continue the drilling. The Farmout Agreement, from the standpoint of the Farmee, should thus contain provisions permitting the Farmee to cease to drill the earning well, and possibly still to earn, when such conditions are encountered.

Of course, the Farmor could always agree to "let the Farmee off the hook," even though inadequate excuse provisions were contained in the Farmout Agreement. Often, a reasonable Farmor will do just that. However, despite the general commonality of interest between the Farmor and the Farmee, (see § **3.03(b)[1]**), conflicts may exist between them due to the differing nature of their interests (one cost-bearing, the other initially cost-free) and possible partially-differing objectives. The Farmor may have additional acreage located in the area which might be better located on the structure, or which might be worth drilling if further information were gathered as to which techniques did and did not appear to work in dealing with the drilling conditions encountered by the earning well. Under these circumstances, a Farmor might wish the drilling of the earning well to continue under circumstances where such a continuance would not be in the Farmee's best interests. In any event, most Farmees would (and should!) feel somewhat uncomfortable in relying upon the good graces of their Farmors as a safety net against potentially disastrous continued drilling.

The industry recession that commenced in 1982 has emphasized the significance of "excuse" provisions in the Farmout and the potentially conflicting positions of the Farmor and the Farmee. With his limited budget, the Farmee was in a much poorer position to live with cost overruns which might have been acceptable when drilling dollars were more readily available. Likewise, if preliminary indications were disappointing, he might prefer to discontinue the drilling of the well rather than spend any more of his limited drilling dollars on a now-questionable project. However, from the Farmor's standpoint, as his own budget also contracted—making "own account" drilling less available as a method of keeping his leases alive—and as candidates for Farmouts become more scarce, this "generous" attitude toward his Farmee might equate to a permanent loss of his leases.

Thus, the Farmee would prefer *very liberal* excuse provisions. The Farmor would prefer that the escape clauses were *much tighter*, leaving it to be in the Farmor's discretion as to whether or not he would be more liberal, in practice, in releasing the Farmee from his obligations or in permitting him at least a partial earning.

Older Farmout Agreements normally excused the Farmee from further drilling if "granite or other impenetrable substances" were encountered. However, there are conditions other than impenetrable substances which could literally prevent the drilling of an earning well to the Contract Depth. Accordingly, this provision was often expanded to excuse further drilling when the Farmee encountered "other conditions in the hole making further drilling *impossible*."

While this expansion of the excuse conditions is certainly desirable from the standpoint of the Farmee, it does not protect him against many conditions under which further drilling on his part hardly seems justified. For instance, what if he were still able to "make hole," but only at such a cost that would make it idiotic to continue to drill? To protect the Farmee under these circumstances, the better-drafted (from the standpoint of the Farmee) Farmout Agreements today provide that further drilling will be excused when the Farmee encounters conditions which make further drilling *impractical* (or impracticable).[37]

While it is relatively easy to determine whether or not the drilling of an additional well is "impossible," the term "impractical" is much more vague; and reasonable (or unreasonable) parties could disagree as to its interpretation. For this reason, some Farmout Agreements attempt to define the conditions which will constitute impracticality. One such provision is the "72 hour continuous efforts" clause, which permits the Farmee to cease further drilling if he has made a good faith, continuous drilling effort for 72 hours without results. However, many Farmors are unwilling to accept such a provision, since the fact of whether or not the drilling effort has been continued "in good faith" for 72 hours may be somewhat difficult to establish. Likewise, from the standpoint of the Farmee, if some, but relatively little, progress is being made, the relief provided by the 72-hour clause may prove illusory.

Another provision, frequently used in the past in Canada to afford relief to the Farmee, was to provide that further drilling was

---

[37] *Cf.* Article VI. A. of the AAPL Form 610 Model Form Operating Agreement (1982 version). Few Farmors will object to such a provision, since they will also want it they are Farmee.

excused *"if, in the opinion of the Farmee,* further drilling is impractical." While such a provision is quite acceptable from the Farmee's standpoint, it would be unacceptable to most Farmors, since it would leave the determination of impracticality almost solely to the discretion of the Farmee. Even the addition of such "qualifiers" as "in good faith" or "in accordance with the customary industry operating practices" would leave the parties in almost the same position as to uncertainty of application.[38]

Despite the many gray areas which could arise in interpreting an "impracticality" provision, and despite the caveats given in the **Introduction** to this book concerning the dangers in today's industry environment of relying on the "good faith" or "reasonableness" of a party to an oil and gas agreement, in practice there have been relatively few problems under the generalized "impracticality" excuse provision. However, under certain circumstances, a Farmee might desire additional protection. For instance:

1. What if further drilling proves *unexpectedly expensive*, but not necessarily "impractical?" (The existence of these conditions will frequently be encountered in drilling an initial well in a "wildcat area.")

2. What if a well *threatened to blow out*, or encountered conditions which *endangered potentially-productive upper horizons?* Would this constitute "impracticality?"

3. What if the well appeared to be *poorly located on the structure, or it otherwise appeared unlikely that favorable results would be encountered* if the well were drilled to total depth? Would this constitute "impracticality?"

The industry has normally interpreted "impracticality" as excusing further drilling only under those circumstances where such

---

[38] As to the use of such qualifiers in discretionary situations, see § **2.04(b)** *et seq.*

drilling was not impossible, but only an idiot would continue to drill. Accordingly, in Canada, the term "inadvisable" is frequently substituted for "impractical" as an excuse for further drilling. Certainly, this is more favorable from the standpoint of the Farmee. Here, again, a Farmor may be unwilling to accept this provision: it will be a question of how badly he wants to make the Farmout and how critical information concerning "deep" conditions is as an objective of his Farmout.[39]

The Farmor may be willing to be more lenient in the defining of "excuse" conditions if the Farmee agrees to the drilling of a mandatory Substitute Well. However, the mere fact that the Farmout Agreement contains a mandatory Substitute Well provision should not always serve as grounds for significantly liberalizing the excuse provision. See § § **3.03(b)[4][d]** and **3.03(b)[4][e]**.

On broad-area exploratory Farmouts, another form of "excuse" provision may be to convert certain otherwise "mandatory" wells to "optional" wells unless a specified minimum number of the wells initially drilled are productive or establish production which meets a certain minimum standard. See § **3.03(b)[3]**.

In areas where significant cost overruns appear a real possibility, a new excuse clause is sometimes appearing in Farmout Agreements. Under this excuse clause variation, the Farmee is authorized to cease further drilling at such time as it has expended *a specified number of dollars* in drilling the earning well. Normally this figure will be anywhere from 150% to 200% of the estimated cost of drilling the earning well. Under an alternative, the Farmee is not excused from continuing to drill the well once the specified dollar amount has been expended, but the Farmor is now required to pick up a portion of the drilling costs from that point on if the drilling of the well is

---

39 As to the importance of varying Farmout objectives on the structure of the deal, see § **3.01(a)**.

continued.[40]

Other variations of the "dollar amount" excuse/earning clause would include the following:

1. The Farmee *still earns, but only to the depth drilled.*

2. The Farmor is now *obligated to make a Dry Hole or Bottom Hole Contribution* toward the further drilling of the well.[41]

3. The Farmor would now be given a *"Take-over" Right*, with a possible addition that the Farmee might back into an interest in the well after Payout (or a multiple Payout) to the Farmor.

The Farmor may also wish to consider limiting the dollar-amount credit to expenditures not involving any negligence (or at least gross negligence or wanton or willful misconduct) on the part of the Farmor.[42]

The dollar limitation provision has its own pitfalls; the contract must clearly specify how such expenditures are to be calculated. A frequent provision is to provide for a measurement based upon an attached "COPAS" accounting procedure. Likewise, on an extremely deep and time-consuming test, the Farmor may wish to provide for adjustment of the dollar amount based on a specified cost index.[43]

[40] It is extremely important, in drafting the Agreement, to clearly specify whether the Farmor *must* pick up such percentage of the costs if the Farmee elects to continue drilling the well, or merely must pay this percentage of the costs if the Farmor insists that the Farmee continue to drill.

As to similar dollar amount earning provisions as a substitute for a multiple well obligation, see **§ 3.03(b)[6][c]**.

[41] Such a support provision could be *mandatory* on the Farmor *if the Farmee chose to continue the drilling;* or only a *condition* of any further drilling obligation.

[42] For similar limiting provisions in computing Payout, see **§ 3.04(b)[2][c-4-b]**.

[43] As to similar problems of measurement, with and without cross reference to COPAS, see **Chapters Four** and **Eight**. As to accounting procedures in general see **Chapter Four**.

However, the dollar limitation excuse clause does offer a potentially viable solution to problems of cost overruns in rank wildcat areas where drilling may not be impossible or impractical, but simply highly more expensive than originally anticipated.

The "cost limit" excuse provision should be used as an *additional* safety valve for the Farmee and not as the sole grounds for ceasing the drilling. If further drilling is impossible, impractical, or inadvisable, the Farmee should not be required to continue his efforts merely because a given dollar amount has not as yet been expended. Whether the excuse of further drilling will be based upon "impracticality," some specified tests as outlined above, or the more liberal "inadvisable" provision is simply a point for negotiation between the parties.

§ 3.03(b)[4][d] "Substitute Well" Provisions. Often, either as a condition of being excused from a mandatory drilling obligation (see § § **3.03(b)[3]** and **3.03(b)[4][c]**), or as a condition of earning, the Farmee will be required to drill a *Substitute Well* at a specified location.

Substitute Well provisions can be advantageous for both the Farmor and the Farmee. From the Farmor's standpoint, a Substitute Well requirement and/or condition assures (or assists in securing) the evaluation of the formation which was the objective of the Farmout. From the Farmee's standpoint, a provision for a Substitute Well may make the Farmor more apt to let the Farmee "off the hook" on an initial mandatory well (see § **3.03(b)[4][c]**), or will permit the Farmee to earn where the original earning well could not be drilled to the required Contract Depth and such drilling is a condition of earning (see § **3.03(b)[4][e]**).

Under the typical Farmout arrangement, the Farmor will reserve a relatively small interes in earning well production until the Farmee has attained *Payout.* (See § § **3.01(b)[4]** and **3.04(b)[3]** *et seq.*) Where a Substitute Well is provided for in the Farmout Agreement, the Payout provision must clearly specify whether the Farmee will be entitled to recover his *aggregate costs* of drilling both the initial and Substitute earning well, or the costs of the Substitute Well alone, before the Farmor is entitled to convert or increase his pre-payout interest. (Most Farmout Agreements are hopelessly silent or unclear in this respect, or reach the result through a literal application of contract wording which was not drafted with this problem in mind.)

There is no clear industry custom as to which of these cost-recovery practices would be looked at as "standard"; thus, whether or not the costs of the original well will be included in the Payout computation in addition to the Substitute Well costs is simply a matter for negotiation. However, irrespective of how this matter is resolved, the Farmout Agreement should clearly specify the method of computing Payout.[44]

Another area of negotiation between the Farmor and the Farmee concerning any Substitute Well provision is whether the drilling of such a well, when it is a substitute for a mandatory well, is similarly to be mandatory on the Farmee, or merely a condition of the earning. This can be a serious consideration from the standpoint of the Farmee. Not only will the Substitute Well obligation increase the Farmee's costs (which may already be in the "overrun" category because of the conditions which led to the abandonment of the initial well), the Farmee may also find that the same conditions which made it impossible/impractical, etc., to drill the initial well to the Contract Depth may extend throughout the Farmout Area and similarly prevent the drilling of any Substitute Well to the specified depth.

From the standpoint of the Farmor, the same considerations that led him to insist that the initial earning well be mandatory normally will make him prefer that the Substitute Well also be a mandatory obligation on the Farmee. From the Farmee's standpoint, the dangers that the attempted drilling of the Substitute Well to the Contract Depth will encounter the same subsurface conditions are also very real. A midground solution would be to provide that the

[44] As to contractually defining "Payout," see **Chapter Eight**.

Farmee *shall* drill the Substitute Well (i.e., that the Substitute Well will be mandatory) *unless there are reasonable grounds to believe that similar conditions will exist in connection with the Substitute Well.*

The parties should likewise consider, in connection with any Substitute Well provision, what provisions should be made for the *location* of the Substitute Well (see § **3.03(b)[2]**), and what provisions should be made for an extension (if any) of the *Completion Date* requirements (see § **3.03(b)[1]**).

If the Farmout Agreement provides for a mandatory Substitute Well, it may appear on first consideration that the Farmor will not really care how broad any "excuse" provisions may be concerning the drilling of the initial well to the Contract Depth. (See § **3.03(b)[4][c]**.) However, time considerations may become a problem if a Substitute Well must be drilled;[45] likewise, where initial well costs are also to be taken into account in determining Payout, it is essential, from the Farmor's standpoint, that no Substitute Well be drilled unless it is absolutely necessary. As if no mandatory Substitute Well were provided for.

§ 3.03(b)[4][e] Effect on Earning. The mere fact that a Farmee is excused from further drilling due to conditions encountered in the drilling of the initial earning well does not necessarily mean that the Farmee will still earn his rights (or the same rights that he would otherwise have earned) under the Farmout Agreement. The fact of *any* earning, or the *extent* of that earning, is still a subject for negotiation.

There are no industry standards as to whether or not a Farmee will earn, or the extent to which he will earn, in the event that continued drilling of an earning well is excused. There are as many

---

[45] The Completion Date specified in the Farmout Agreement usually represents the *latest date* by which the Farmor wishes an earning well to be completed if the Farmee is still to earn. In most exploratory Farmouts, an earlier completion is far preferable, since this allows further time to evaluate well results and to conduct additional drilling before leases begin to expire.

possible solutions and alternate provisions concerning the impact of a justifiable cessation of drilling as there are Farmout Agreements. Possible provisions include the following:

1. The Farmee *still earns, without depth or acreage limitations* (or with only such limitations as would have applied if the earning well had been drilled to the specified Contract Depth).[46]

2. The Farmee *still earns, without depth or acreage limitations* (or with only such limitations as would have applied if the earning well had been drilled to the specified Contract Depth), *so long as the earning well establishes production at any depth.*

3. The Farmee *still earns throughout the Farmout Area* (or such portion of the Farmout Area as the earning well would have earned), *to the depth drilled only.*[47]

4. *The Farmee still earns throughout the Farmout Area* (or such portion of the Farmout Area as the earning well would have earned) *to the base, etc., of the deepest productive formation penetrated.*

5. A variation of any of provisions "1" through "4" which *limits the earning to the earning-well Drilling and Spacing Unit.*

6. The Farmee *still earns, but* **only if** *a Substitute Well is drilled.*

7. The Farmee *still earns, but* **must** *drill a Substitute Well.*

---

[46] As to the acreage earned by a particular well, see § § **3.03(b)[6]** *et seq.* and **3.04(b)[4]**.

[47] As to depth limitations on earning, see § **3.03(b)[5]** *et seq.*

Under certain circumstances, a Farmee may be reluctant to continue the drilling of an earning well which has established apparent production in upper zones if the continued drilling might endanger the hole or the formation due to excessive pressures, threats of salt water intrusion, etc. Under these circumstances, if the Farmor is equally insistent that he wishes to see the deeper formations evaluated, an impasse may arise. A reasonable mid-ground might be to provide that continued drilling of the initial well would be excused if the Farmee establishes commercial production at a lesser depth, *so long as the Farmee commences a new test to the Contract Depth within a specified period of time.*

In such a provision, two apparently minor variances in wording could lead to significantly differing results. If the contract provides that the Farmee earns to the lesser depth, but is *obligated* to drill the additional test, the Farmor's only remedy if the additional well were not drilled would be to sue for damages. However, if the provision read that the Farmee would earn in the first well only *so long as* the additional test well were drilled, the drilling of the second well becomes a *condition* of the earning: the Farmee would lose his rights in the initial well unless the second, deeper tests were drilled.

It must be noted that, each time that a Farmee is permitted to earn, at least in part, without *any* well having been drilled to the desired Contract Depth, the Farmor will now have acquired a "partner" who may be reluctant to participate in deeper drilling, and whose presence will complicate the negotiation of any new deep-test Farmout with some third party.

§ 3.03(b)[5] Depth Earned ("Farmout Depth") Provisions

§ 3.03(b)[5][a] In General. Under certain circumstances, the Farmee will earn *without depth limitation* so long as the earning wells

---

[48] As to whether such conditions would constitute "impracticality" of continuing, see **§ 3.03(b)[4][c]**.

are drilled to the specified minimum Contract Depth. Under other circumstances, the Farmee will earn to a *specified depth* by drilling to the Contract Depth and cannot increase his rights by drilling deeper.[49] However, more common provisions are: (i) *to limit the depth* earned by the Farmee *to a depth which bears some relationship to the depth drilled* in the earning well; and (ii) to permit the Farmee *to increase the depth* to which he earns *by voluntarily drilling below* the specified minimum *Contract Depth.*

Depth Earned provisions might be looked at as more logically falling in the category of *Interests Assigned and Reserved.*[50] However, due to the relationship between "minimum depth" (*Contract Depth*) and "depths earned" (*Farmout Depth*) provisions, the subject will be dealt with at this point.

§ 3.03(b)[5][b] Pitfalls in Defining Depths Earned. It is easier to state the concept (or to make a brief reference in a loosely-drafted letter) to limit depths earned to depths drilled than it is to draft an earning provision which will be fair both to the Farmor and to the Farmee.

From the standpoint of the Farmor, he will wish the Farmee to earn no deeper than the depths to which he has "paid the price tag," i.e., the depths to which he has actually drilled, thus incurring both (and establishing) the costs and the drilling conditions that will be encountered in reaching such a depth. From the standpoint of the Farmee, however, he will not wish to be "cheated" out of his rights to a given formation whose productivity he has established, merely because in a given portion of the Farmout Area, a relatively slight "tilting" of the formation means that it is now encountered at slightly lower depths than the depths at which the formation was present in

---

[49] As to provisions limiting the *maximum* depth to which the Farmee can earn, see § § **3.01(b)[2]** and **3.03(b)[5][d]**.

[50] As to such provisions, see § **3.04** *et seq.*

the earning wells.[51]

Initial drafting attempts limited the Farmee's earning to a specified depth, e.g. 10,000 feet below sea level, or "to the depth drilled" in the earning well.[52] But, from the standpoint of both parties, this could now mean that a given formation could be sliced in half; and from the standpoint of the Farmee, could divest the Farmee of his rights in portions of the Farmout Area to a formation which—in fairness, though possibly not under the wording of the Farmout—he should have earned. See **Figure 19**.

This latter consequence—loss of *any interest* in the tested formation—becomes anywhere from a very real possibility to a certainty under an "earn to a specified depth/depth drilled" earning provision if the Farmout Area is fairly extensive. Variations of this provision, such as permitting the Farmee to earn 50 to 100 feet below the depth drilled, are not much better if the Farmout Area is extensive: the "safe harbor" will either be so limited (at the insistence of the Farmor) that it might offer relatively little protection to the Farmee, or so large (at the insistence of the Farmee) that it may permit the Farmee to earn rights and formation for which he has paid an adequate drilling and testing "price."

At this point, the lawyers entered the scene to propose what, from a legal standpoint, seemed the perfect solution — the Farmee would earn to the *base* of a specified formation, or, in the alternative, to the base of the deepest formation (or productive formation) penetrated.

---

51 The impact of such a tilting is illustrated in **Figure 19**.

52 For all effects and purposes, the legal consequences of specifying an *exact depth* to which the Farmee would earn his rights, or limiting the earning to the *depth drilled*, are the same: in both cases, once the earning well has been drilled, the Farmee's rights will now be limited to an exact subsurface or subsea cutoff point.

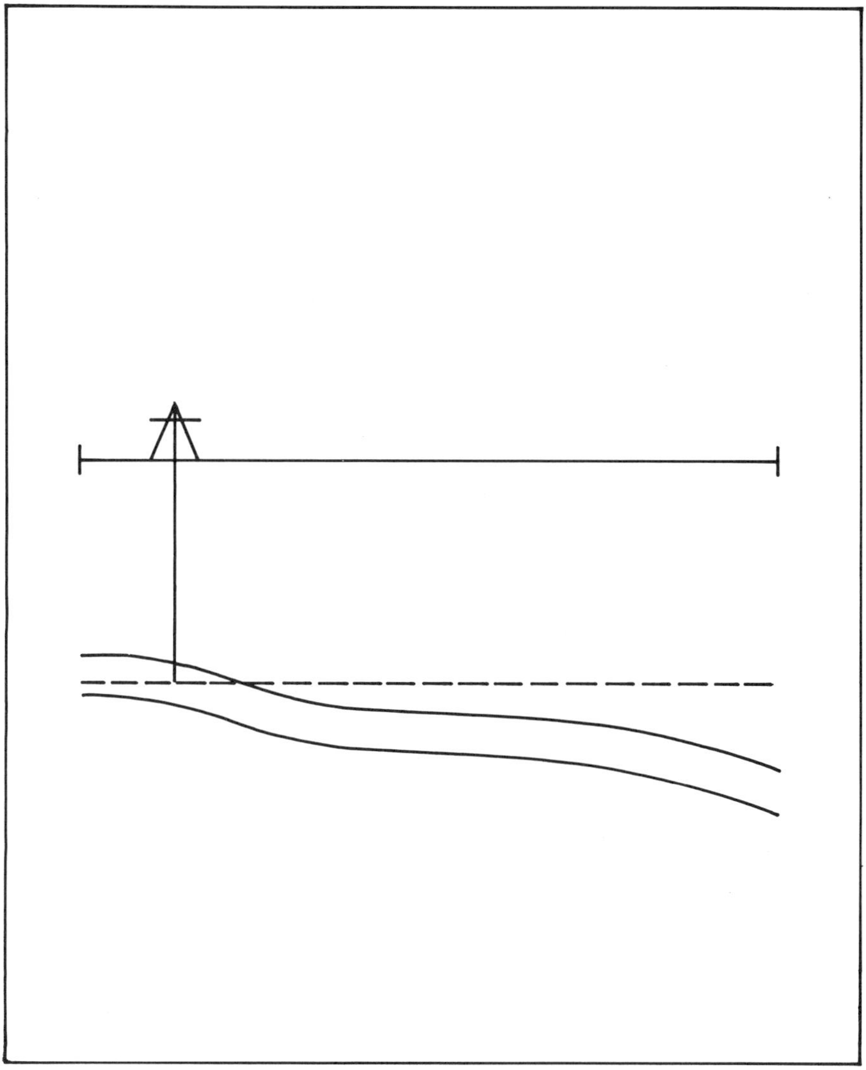

Figure 19: Impact of Formation Tilting on "Earn to Depth Drilled" Provision

While this provision did have the advantage of certainty,[53] it could lead to a Farmee earning rights far below the depths actually attained and evaluated by the earning well in the event of faulting or folding of a formation in a multilense situation, or when "thrusting" is encountered. Thus, as illustrated in **Figure 20**, a Farmee, by drilling to point "A," would earn rights to point "B" when such conditions were encountered, despite the fact that point "B" lay at a depth far below the depth actually drilled in the earning well. On the other hand, if the formation had pinched out or eroded away in a given part of the Farmout Area, there would be a serious question as to whether the Farmee had earned *anything* in that portion of the Farmout Area.

In an attempt to eliminate the pitfalls encountered in drafting other Depth Earned provisions, Farmout Agreements today customarily provide that the Farmee will earn to the *"Stratigraphic Equivalent"* of the depth drilled. However, despite a one-time almost universal use of the Stratigraphic Equivalent provision, this common "solution" to the depth earned problem can create many pitfalls of its own. See § **3.03(b)[5][c]**.

§ 3.03(b)[5][c] Pitfalls in "Stratigraphic Equivalent" Earning Provisions. There are two reasons for the widespread use of the Stratigraphic Equivalent earning provision. The first is that it *does* solve the "tilting" problem mentioned in § **3.03(b)[5][b]**. The second "reason" is the fact that the term *"Stratigraphic Equivalent"* has a reassuring (but misleading) sound of certainty; thus, non-geological professionals frequently use the term with no understanding of its true meaning, but with a comfortable feeling that "nothing that

---

53 *Hopefully*, the test provided certainty. This was true, however, only if an *established geological marker* was selected — one on which there was little disagreement among geologists as to the nomenclature or what sands actually constituted the formation — and there was no difficulty in agreeing that a sand in one part of the Farmout Area was the same sand encountered in the earning well. Unfortunately, this latter issue was one upon which equally competent geologists might disagree if there had been an interruption of the formation within the Farmout Area.

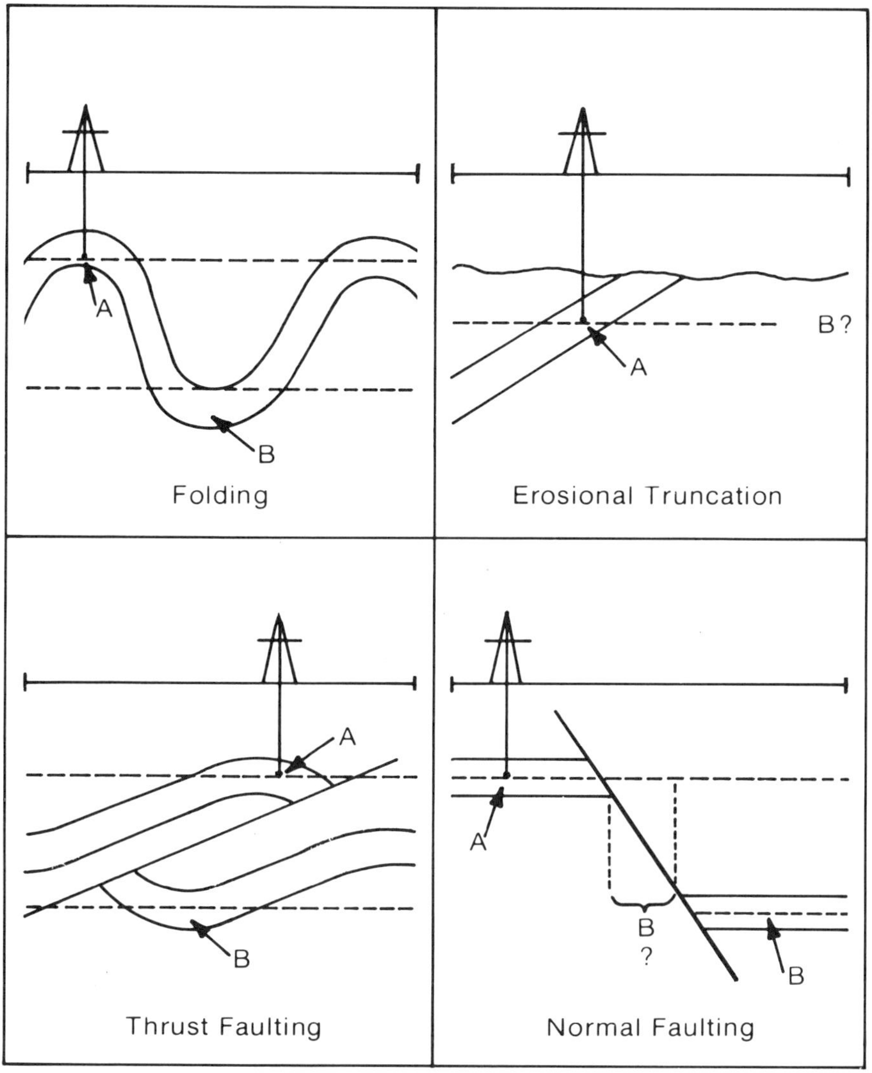

Figure 20: Impact of Various Subsurface Conditions on "Earn to Base" and "Stratigraphic Equivalent" Provisions

sounds that good could fail to solve the problem."

To the geologist, Stratigraphic Equivalent refers to the point in *geologic time* when the formation was laid down.[54] Thus, over an extensive Farmout Area, it could refer to a formation which was deeper (or shallower) than the formation actually penetrated by the earning well. Likewise, while the Stratigraphic Equivalent provision *does* solve the "tilting" problem illustrated in **Figure 19**, it *does not* solve the faulting, folding, and other problems indicated or illustrated in **Figure 20**. In other words, in each one of these cases, the Farmee would still earn to point "B", even though the earning well was only drilled to point "A".[55]

If it is not anticipated that any unusual conditions of the types illustrated in **Figure 20** will be encountered in the Farmout Area, a Stratigraphic Equivalent earning provision may be appropriate.[56] However, if there is any reasonable possibility that such conditions might be encountered, a better provision might be that the Farmee would earn to a *Farmout Depth* defined as follows: " 'Farmout Depth' shall mean 100 feet below the depth drilled in the earning well;[57] provided that, if the Stratigraphic Equivalent of the depth drilled in the earning well exceeds such depth, 'Farmout Depth'

---

54 *"Stratigraphic Equivalent"* has several geological meanings; the one just given — *"temporal Stratigraphic Equivalency"* — is the one which would normally be applied in interpreting the Farmout Earning provision. The other common meaning — *"Lithologic Stratigraphic Equivalency"* (the "same rock") — creates many of the same problems.

55 Confusion can also arise if a formation varies in thickness throughout the Farmout Area. For instance, if a Farmee drilled 50 feet into a formation which was 100 feet thick at the earning location, but the formation was 400 feet thick in other portions of the Farmout Area, would the Farmee have earned the entire formation, 50 feet into the formation, or 200 feet into the formation, at those locations with 400 feet of sand? And, what if the formation was *4,000* feet thick (the deeper portions never having been tested)?

56 But, in a rank wildcat Farmout, how sure can even the geologist be that such conditions will not be encountered?

57 This protects the Farmee if the formation pinches out, erodes away, etc.

shall mean such deeper depth, but in no event more than ____ feet below the depth drilled in the earning well."[58]

§ 3.03(b)[5][d] Restricting the Farmout Depth and "Sandwich" Farmouts. As mentioned in § **3.01(b)[2]**, the Farmor may decide to limit the formations which may be acquired by the Farmee irrespective of the Farmee's willingness to drill to certain depths. The Farmor may determine that certain shallow formations, which are "proven" in nature, should not be committed to the Farmout.[59] The Farmor may also determine that certain "deep" zones have such potential that he does not wish the Farmee to earn rights to these zones even though the Farmee may be willing to test them.

If the Farmor elects to reserve certain shallow rights, and further elects, as is common, to limit the Farmee's earning to depths drilled (or elects to reserve certain deeper zones from any commitment to the Farmout), a *"sandwich" Farmout* will result — the Farmee will earn certain intermediate zones which lie between shallower and deeper rights reserved by the Farmor. It is critical in such a sandwich Farmout that the Farmor *clearly specify his intent* to reserve the upper horizons, since courts tend to interpret a Farmout in accordance with the more common intention of assigning to the Farmee all rights from the surface of the earth to the depths earned.[60]

An excellent way of clearly indicating the intention of creating a sandwich Farmout is for the Farmor *expressly to reserve* both the

---

[58] This limits the Farmee to the *lesser* of the Stratigraphic Equivalent or the final "cut-off" point.

[59] Despite the "proven" nature of certain upper formations, the Farmor more frequently elects to commit them to the Farmout as an incentive for the Farmee's deeper drilling commitment.

[60] In one case, the Oklahoma Supreme Court held that an agreement to assign leases "to the Humphrey Sand only" showed an intention to assign the leases from the surface to the base of the Humphrey Sand. See *Waldrip v. Hamon*, 136 F. Supp.

As to the necessity for clearly stating any "Formation" designation, see § § **3.03(a)** and **3.03(b)[5][b]**.

upper and lower formations, and to attach a copy of an electric log from a prior well as an exhibit to the Farmout, indicating on such log the formations available for earning by the Farmee, and the formations to be reserved by the Farmor.

Where it is unclear exactly to which depths the Farmee will earn by the drilling of the earning wells, since that earning will be dependent upon the depths actually drilled (see § **3.03(b)[5][a]**), the Farmee's earning, as discussed above, will normally be tied into some variation of a *Depths Drilled* provision. See § **§3.03(b)[5][a]-3.03(b)[5][c]**. However, the mere fact that the Farmor has specified a *Contract Depth* in the Farmout Agreement will not limit the depths earned by the Farmee where a *Depths Drilled* earning provision is used: the Contract Depth is a *minimum depth*, and, as such, imposes no restraints on the depths to which the Farmee can earn. If the Farmor intends to restrict the maximum depth to which the Farmee can earn under the Farmout Agreement, this *Farmout Depth* restriction must be clearly stated in the Farmout Agreement. Accordingly, a provision that the Farmee *will in no event earn below the top of a specified formation* (or below a specified depth) is necessary where the maximum earning is to be so limited.

Where ownership of deep and shallow rights is not identical, the owner of the shallower rights should insist on an indemnification provision to protect against damage to his formations when the owner of the deeper rights drills through these zones, and should seek a *Take-Over* provision if the well is not productive in the zones owned by the drilling party, but is potentially productive in zones owned by the holder of the shallower rights.[61]

---

[61] The drilling party is normally reimbursed only for the costs of drilling the well to the zones in which the non-drilling party wishes to attempt to complete the well (or to the base of the rights owned by the non-drilling party), and is sometimes limited to reimbursement for the costs of plugging back to the shallower formation. Allocation problems in properly arriving at such costs will be discussed in **Chapter Four**.

§ 3.03(b)[6] Acreage Earned and Multiple Earning Well Provisions.

§ 3.03(b)[6][a] In General. In the developmental Farmout, a Farmee customarily earns an interest only in those Drilling and Spacing Units on which earning wells are drilled. However, in the traditional exploratory Farmout, risk-adjusted economics (and the bargaining position of the Farmee) normally dictated that, in addition to earning an interest in the earning-well DSU, the Farmee would earn an additional interest in certain surrounding acreage, referred to as *excess* or *Outside Acreage.*

The amount of the Outside Acreage earned by the drilling of each earning well and the size of the interest assigned in such acreage would vary according to regional practices, the costs, risks, and upside potential involved, and the bargaining position of the parties.[62] As little as one additional DSU, or as much as all eight direct and diagonal offsets (or even more extensive interests in broad-area exploratory arrangements) might be earned. In some areas, such as Southern Louisiana, there might even be no Outside Acreage earned—earning would still be limited to the earning-well DSU, even in an exploratory Farmout. Customarily, however, an interest in Outside Acreage—normally in the form of an undivided Working Interest in such acreage—was assigned as an added incentive in most exploratory Farmouts.[63]

In the 1977-1981 period, drilling dollars were so plentiful, acreage so tight, and economics apparently so favorable (based on that period's erroneous assumptions concerning short term price and demand conditions), that Outside Acreage as an exploratory Farmout "sweetener" tended to disappear, except in the Rocky Mountain area.

---

62 Traditionally, the Farmor would reserve an interest in such acreage equal to his convertible (Reversionary) Working Interest in the earning-well DSU.

63 As to the use of "checkerboard" arrangements as an alternative to assigning and reserving undivided Working Interests, see § **3.04(b)[3][c]**.

However, with the return of "normal" (and *subnormal*) conditions in 1982, Outside Acreage returned as a characteristic of the exploratory Farmout in most parts of the country.

An extensive *Farmout Area* cannot be thoroughly evaluated by the drilling of a single well. The Farmor thus has several choices. He can limit the amount of acreage committed to the Farmout (see § § **3.01(b)[2]**, and **3.02(b)**), or he can commit more acreage to the Farmout, subject to a *multi-well drilling commitment* on the part of the Farmee.

As discussed in the foregoing sections, there are certain advantages to limiting the size of the Farmout Area, even though this will reduce the extent of the commitment which the Farmor can secure from the Farmee. A less extensive Farmout Area will permit the Farmor to retain additional acreage for drilling solely for his own account, or for farming out to other parties. Simultaneous Farmouts are other alternatives which will permit a more rapid development of the Farmout Area than could normally be secured from a single Farmee.[64]

However, in a desirable Farmout Area, the Farmor can secure a much greater commitment from the Farmee if he will commit substantial acreage to the Farmout; and the commitment which the Farmor can negotiate will normally be much larger than "the sum of the parts" of commitments that could be received under multiple Farmouts, each of which involved relatively small amounts of acreage.

As a result, the Farmor must balance his desire for flexibility with the more thorough, normally-cost free, benefits of an extensive drilling and testing undertaking by the Farmee.

---

[64] The Farmout Agreement could and sometimes does provide for the *simultaneous commencement* of *several earning wells* by the Farmee; however, such a provision is unusual, and difficult to negotiate, and carries a high price in easing other earning requirements, or substantially sweetening the deal, to the Farmee.

The number of wells to which a Farmee will commit a given area will be a subject of negotiation. Likewise, in *optional* earning situations,[65] the amount of acreage earned per well drilled, as mentioned above, will depend upon regional practices and the relative bargaining positions of the parties. Thus, as mentioned, in some areas, e.g., Southern Louisiana, even a high-risk well frequently earns the Farmee only a single Drilling and Spacing Unit. However, as just discussed, in most parts of the United States, in an exploratory Farmout situation, a Farmee will earn an interest in undrilled offsetting Drilling and Spacing Units by the drilling of each earning well.

In Canada, complicated after-tax computations are made to arrive at the acreage to be earned by each earning well by contrasting the cost of directly purchasing equivalent acreage with the estimated cost of "buying" such acreage through bearing the cost of drilling. In the United States, the determination of acreage earned by each well drilled is more commonly based upon regional practices and the expected costs and risks of the drilling.[66]

In a developmental Farmout, the Farmee continues to earn only a Drilling and Spacing Unit at a time. This normally leads to a *Continuous Drilling* provision (see § **3.03(b)[6][b]**), unless the Farmout is to be limited to a single Drilling and Spacing Unit.

§ 3.03(b)[6][b] "Continuous Drilling" Provisions. In a multi-well developmental Farmout, the Farmout Agreement will normally provide that the Farmee will earn on a Drilling-and-Spacing-Unit by Drilling-and-Spacing-Unit basis only. The Agreement may call for an assignment to the Farmee of each Drilling and Spacing Unit as drilled, or may provide for an assignment of rights throughout the

---

[65] As to the *mandatory* versus *optional* nature of the earning wells, see § § **3.03(b)[3]** and **3.03(b)[4]** and its subsections.

[66] Obviously the relative position of the parties as to budget constraints or lacks thereof, lease expiration dates, rig availability, etc., also enter into the negotiations.

entire Farmout Area, subject to a duty to reassign (or an automatic termination) as to any undrilled locations.[67]

The necessity for continuous drilling obligations is becoming more significant as leases are characterized by shorter primary terms, express development provisions, and "Pugh" clauses.[68] The emergence of "Top Leasing" as a common practice by many oil companies has also led to the necessity for more rapid development of the Farmout Area.[69]

The "continuous drilling" will normally require, as a condition of further earning, that a new well be commenced within a specified period of time following the completion of the preceding well. In the developmental Farmout, the grace period is normally 30 to 90 days.

From the standpoint of the Farmee, it is critical that the time allowed between the date of completion of the present well and the commencement of the next well be sufficient for completing title work and making necessary rig arrangements. The Farmout Agreement should likewise clearly define the terms "commence" and "complete;" otherwise, a Farmee could inadvertently lose his rights (or good faith disagreements between the Farmor and the Farmee could arise) because of uncertainty or disagreement concerning the meaning of these terms.[70]

In an exploratory Farmout, you may see a variation of the *Continuous Drilling* provision, where further earning *by block* is based upon a similar obligation. However, in the exploratory

---

[67] As to *Present Assignment* versus *Assign when Earned* approaches, see **§ 3.04(b)[2][c-5-a]**.

[68] The "Pugh" clause is a provision in the oil and gas lease providing that the lease will terminate either at the end of the primary term or as to all undrilled locations. Certain states have also enacted statutory "Pugh" provisions.

[69] A "Top Lease" is a lease taken by a second Lessee while an earlier lease in favor of a prior Lessee is still in effect. Such leases should be contrasted with "Extension" or "Renewal" Leases, which are leases taken by an existing Lessee to extend *his* rights for an additional period.

[70] As to possible definition of "commence" and "complete," see **Chapter Eight.**

Farmout, it is important to provide that only the drilling of a well which would qualify as an *earning well* continues the Farmee's right to earn. See § **3.03(b)[6][c]**.

As to incentives to encourage a Farmee to drill the maximum number of earning wells, even where such drilling is *optional*, see § **3.03(b)[6][c]**.

§ 3.03(b)[6][c] Multiple Earning Well Provisions in Exploratory Farmouts. As discussed in § **3.03(b)[6][a]**, if the Farmor is willing to commit a significant amount of acreage to an exploratory Farmout, he will not want the Farmee to earn an interest throughout the entire Farmout Area by the drilling of a single earning well, particularly when the Farmout Area covers a number of potential prospects.

If the Farmout Area covers anywhere from 6,000 to 12,000 acres, the Farmout Agreement will most commonly provide that the Farmout Area will be divided into "Blocks," with the Farmee to earn an interest in each block on which an earning well is drilled. However, if the Farmout Area is even more extensive, covering tens to hundreds of thousands of acres, it will be difficult to pre-designate the "Blocks." Under these circumstances, a more common provision is to provide that the Farmee will earn a specified amount of acreage around each earning well drilled. Here, "sweeteners" to increase the amount of acreage and/or depths earned by the drilling of each earning well, if a large amount of *optional* drilling is undertaken by the Farmee, are quite common. (See § **3.04(b)[3][d]**.)

Whatever the manner of the earning, and whether all or a part of the earning wells are optional or mandatory, a *Continuous Drilling* obligation will normally be inserted in the Agreement requiring the Farmee to commence the drilling of each new earning well within a specified period of time following the completion of the last earning well. (See § **3.03(b)[6][b]**).

Unlike the developmental Farmout, or even an exploratory Farmout covering a relatively limited amount of acreage or a single prospect, the Farmee will frequently be given a right to *accumulate time* under the broad-area exploratory Farmout. Under such a provision, if the Farmee drills several earning wells in a shorter period than was required to continue the earning (or avoid a breach of a contractual mandatory drilling obligation), he will increase the time available to him before a new earning well must be commenced. Such a provision is advantageous to the Farmee, as it will give him a greater period of time in which to evaluate the results of the earlier wells before additional drilling must be undertaken. However, from the standpoint of the Farmor, there should be a specified *maximum accumulation right* (e.g., 180 days); otherwise, the Farmor may find that a Farmee, by drilling several wells early on in the agreement life, may now wait for a period of months or years before being obligated to undertake further drilling.

In some Farmouts, and if the Farmor is faced with threats of expiring acreage, more than one of the earning wells may be mandatory. (The Farmout Agreement could even require that these mandatory wells be drilled concurrently.) However, by far the more common provision will be to provide that the drilling of all but the first earning well will be optional with the Farmee — a condition of his earning an interest in the particular block.[71]

If the Farmout Area is particularly extensive, and covers multiple potential prospects of a "rank wildcat" nature, the Farmout Agreement may call for a combination of mandatory and optional wells, and may further impose certain seismic and similar obligations on the Farmee. The Farmor will be taking into account that he has tied up potentially attractive acreage; if the Farmee earns an interest in portions of that acreage by a relatively limited amount of drilling, this will greatly restrict the Farmor's flexibility in securing the development of the balance of the acreage. The Farmee, on the other hand, should give considerable thought before

[71] As to the various considerations in designating one or more of the earning wells as optional or mandatory, see § **3.03(b)[3]**.

accepting an extensive "mandatory drilling" obligation, particularly if the acreage is "rank wildcat" in nature.

A characteristic to be prized in a talented exploration geologist is his ability to see prospect potential in areas that other explorationists have overlooked. However, this same desirable characteristic may lead the geologist to be willing to commit to a more extensive *mandatory* drilling obligation than may prove wise. It must be remembered that, no matter how attractive the area, it is still high risk in nature and, under the percentages, will probably not prove productive no matter how geologically intriguing it may have been. Accordingly, it is critical from the standpoint of the Farmee that only a minimum number of mandatory wells be agreed to in such a Farmout, or that the Farmee be permitted to "get off the hook" on further mandatory drilling if the initial wells are disappointing.

The general guidelines concerning acreage to be earned by the drilling of earning wells go out the window where the multi-prospect Farmout having an extensive amount of wildcat acreage is involved. The acreage earned will be dependent upon the degree of risk involved, the estimated costs of the earning wells, the amount of overall acreage being committed to the deal, and the relative bargaining positions of the parties. Thus, a Farmor who is faced with lease expiration and a limited drilling budget will not be able to negotiate as tough "earning conditions" as a Farmor whose leases were still early in their primary terms and who had no such budgetary constraints.

A great deal of flexibility can be exercised in structuring such a Farmout as to the number of wells to be drilled, the acreage to be earned by each, and the mandatory or optional nature of such drilling. The inclusion of a seismic obligation and of "sweeteners" to encourage the Farmee to maximize his drilling is quite common in such arrangements.

For analogous problems in the Exploration Contract, see **Chapter Five**.

Particular problems arise in specifying the location of the various earning wells under a broad-area exploratory Farmout. The general problems involved in designating well locations in multi-well drilling situations has been discussed in § **3.03(b)[2]**. However, under the exploratory Farmout, the Farmor's motivation in committing extensive amounts of acreage over a multi-prospect area is to encourage a *thorough evaluation* of the area. Frequently, the broad-area exploratory Farmout will merely divide the area into blocks, or provide that a given amount of acreage will be earned by the drilling of each earning well. In many such agreements, since it cannot be determined where additional wells should be drilled until seismic and/or drilling has provided more information, the Farmout Agreement may provide that the earning wells can be drilled "at a location of the Farmee's choosing." This would mean that *by drilling multiple wells at the intersection of several earning blocks*, the Farmee could earn an interest in each block (or in the amount of acreage to be earned by each well drilled), even though all but the first of such wells is actually a development well and does not further serve to evaluate the Farmout Area. Likewise, if a commercially-productive well were drilled by a third party adjacent to the Farmout Area, or on acreage within the Farmout Area not owned by the Farmor, the Farmee could earn by drilling a well offsetting such productive well even though this "earning well" was not truly exploratory in nature. While such conduct would not be within the *spirit* of the Agreement, it normally would serve to satisfy the *literal earning obligations* of the contract.

To avoid such an abuse of the earning conditions, the broad-area Farmout Agreement should provide that no earning well may be drilled within a specified distance from any other earning well, or, better still, any other commercially-productive well adjacent to or located on the Farmout Area.[72]

---

[72] If the Farmout is exploratory only as to deeper formations, for the protection of the Farmee the Agreement should provide that a well will still be considered exploratory if it is drilled to (or completed in) a zone not currently producing within the specified distance from such earning well.

A similar "trick" on the part of a Farmor who has earned his rights under some, but not all, of the blocks within the Farmout Area, is to propose a "joint" well at the far edges of one of the earned blocks, directly offsetting an unearned block. The Farmor must now elect either to participate in that well, thus assisting the Farmee in proving up the offsetting, unearned block, or to be carried subject to the stringent "penalty" provisions of the Joint Operating Agreement. (See **Chapter Four**.) Solutions, from the Farmor's standpoint, are to require that any "development" well located within a specified distance of an unearned block must either be drilled at the sole cost and expense of the Farmee, subject to the same reserved overriding royalty and similar rights which apply to earning wells (see **§ 3.04** *et seq.*), or to provide that no development wells can be drilled (at least within the specified distance from an unearned block) until the Farmee has completed the drilling of all earning wells.

The "economics of earning" in a Farmout situation take into account that certain money expenditures by a Farmee should earn certain amounts of acreage. This is usually translated into a *given number of wells*, based upon the *estimated cost of drilling*. However, in the exploratory Farmout, this is just the trouble — no one knows exactly what the cost of the earning wells may be. (This may be a key reason that the Farmor is eager to make the Farmout.)

From the standpoint of the Farmee, a Seismic Option arrangement may prove preferable for earning rights in "rank wildcat" acreage, rather than a "straight" Farmout Arrangement. See **Chapter Five**. However, the Seismic Option doesn't offer complete assurance to the "Optionor" that any wells will actually be drilled on the Contract Area, or provide either party with much guidance concerning what earning wells will actually cost. Perhaps a better answer may be a less conventional, but more practical, earning arrangement for these conditions.

To avoid the dilemma created by cost uncertainties in rank wildcat areas, some Farmout Agreements are now providing that the amount of acreage earned by the Farmee will be determined by the *dollars spent* by him in drilling on the Farmout Area, rather than by the number of wells drilled, e.g., a 2% working interest throughout the Farmout Area for each million dollars expended in specified activities, up to a maximum 50% interest.

Earning by dollars spent does not have to be limited to dollars expended in drilling. If, as is frequently the case, control is lacking in the area, and there is considerable unleased acreage, the Farmee might also receive credit for dollars expended for seismic or additional lease acquisition.

The Farmor may impose restrictions as to the percentage of earning dollars which can be expended in any given portion of the Farmout Area, or for seismic or lease acquisition, and may impose dollar limitations on the amount of credit received for the drilling of any one well.[73] A clear *standard of dollar measurement* must also be established by the Agreement.[74] Subject to these caveats, however, a "dollars expended" earning provision for rank wildcat Farmouts seems an intelligent means of adapting customary Farmout arrangements to the exigencies of today's conditions.

§ 3.03(b)[7] "Productive to Earn" and Other Earning-Well Earning Conditions. In the past, Farmout Agreements normally contained only a *Drill to Earn* earning requirement. In other words, so long as an earning well was drilled to the specified Contract Depth, and the other earning conditions were complied with, the Farmee would earn his rights *even though no effort was made to complete the earning well and the earning well resulted in a dry hole.* Under some Farmout Agreements, the Farmee might be required to make a completion attempt, either on a *mandatory* basis or as a condition of

---

[73] This should vary according to depth.

[74] As to comparable problems under dollar amount "excuse" provisions, see § **3.03(b)[4][c]**.

earning, if the Farmor so requested.[75] Likewise, the typical Farmout would give the Farmor a right, as his option, to take over an earning well which the Farmee elected not to complete.[76] However, *Drill to Earn*, or, at most, *Complete to Earn*, agreements were standard in the industry until recent years.

If a Farmee earns an interest in exploratory Farmout acreage by the drilling of a dry hole, the Farmor may find himself stuck with a potentially reluctant "partner" for developing the balance of the Farmout Area if the Farmee has either been sufficiently discouraged by the initially non-productive drilling or is now out of money. Farmees under such circumstances have been known to apply for an increased spacing pattern to permit them to "back into" an adjacent well previously drilled by the Farmor if their earning well is dry, or to force the Farmor to "back into" the earning well and thus pick up a portion of his cost.[77] Even more significant, however, is the fact that an objective of a Farmout, from the Farmor's standpoint, may have been to expand his own drilling budget by securing a "partner" for additional development of the Farmout Area. The Farmee who has drilled a dry hole may simply now go "Non-Consent" on subsequent wells, defeating this objective of the Farmor.[78] Furthermore, the Farmee will now be entitled to participate in Area of Mutual Interest rights if the Farmout Agreement contains such an "AMI" provision,[79] and must also be consulted concerning delay rental, release, and

---

[75] From the standpoint of the Farmee, any such obligation to complete needed to be qualified by the condition that the demand to complete be "reasonable," "in good faith," or "in accordance with customary and prudent operating conditions in the industry," etc.

[76] See § **3.04(c)**.

[77] To prevent such conduct, the Farmout Agreement should provide that any spacing applications by the Farmee must be concurred in by the Farmor. See § **3.04(c)**.

[78] As to further development of the now jointly-owned "excess" acreage following a Farmout, see **Chapter Four**.

[79] As to Area of Mutual Interest provisions, see **Chapter Seven**.

similar lease administration decisions.

In many instances, a primary objective of the Farmor has been to extend the Farmout Leases beyond their primary terms to protect the Farmor's reserved deep rights. (See § **3.01(a)**.) If the earning well or wells are non-productive, the leases will only be extended for the drilling and any additional "grace" period, and will then expire in the absence of further operations. However, the Farmor's ability to negotiate a new Farmout with a less "burned out" Farmee will be greatly hindered by the fact that the initial Farmee now owns an interest in those leases as a result of his *Drill to Earn* Farmout.[80]

For the same reasons that continuous drilling provisions are common in Farmout arrangements (see § **3.03(b)[6][b]**), Farmout Agreements in many areas are now frequently of the *Productive to Earn* variety. Under such a provision, the Farmee will not earn unless the earning well is not only drilled to the specified contract depth, *but also establishes production.*

Variations on the *Productive to Earn* Farmout are provisions that the Farmee will earn *only to the deepest productive horizon,* or will earn by drilling, but is *obligated to reassign the acreage* if production is not established within the Farmout Area within a specified period (e.g., 12 months).

It is critical from the standpoint of both parties that the term "productive" be clearly defined in the Farmout Agreement. For instance, while a well which is shut-in for lack of a market will not be considered productive in most states for purposes of the Habendum Clause of the oil and gas lease, in the absence of an applicable shut-in royalty clause the Farmout Agreement should specify that such a well will be considered a *Productive Well* for purposes of a *Productive to Earn* provision.

How "productive" is defined should also vary with the objectives of

---

[80] For this reason, some Farmors are now insisting that a Farmee "use best efforts" to renew any expiring Farmout Leases, or notify the Farmor and release the lease from the Farmout some specified period before the lease is to expire if the Farmee doesn't intend to drill it.

the Farmout. If *maintaining the Farmout Leases in effect* is the primary objective, "productive" should be defined as production in quantities to yield a profit over *operating costs*, which would satisfy the requirements of the Habendum Clause. If the objective was *to secure an aggressive drilling partner*, this won't cut it: the Farmee who has drilled a series of expensive, marginally-commercial wells won't be eager to devote much of his remaining drilling budget to further wells in this area. Accordingly, "productive" should here be defined as production which would justify the drilling of an additional well in the area.[81]

From the standpoint of the Farmee under a *Productive to Earn* Farmout, it is critical that he be given the right to continue to drill until he has either established production, and thus earned his rights under the Farmout, or until *he* has decided to throw in the towel. Otherwise, an initial test well might prove up the balance of the Farmout acreage for drilling by the Farmor for his own account, even though the Farmee's test well proved unproductive due to its location on the structure.

§ 3.03(c) Testing and Other Earning Conditions. The information that the Farmor gains from the testing of the Farmout Area may be just as valuable to him as the drilling of the earning wells. Such information will be necessary for the Farmor to determine whether or not he wishes to propose, or participate in, subsequent development or additional exploratory drilling in the balance of the Farmout Area.[82]

---

[81] As to sample definitions of "productive" or "Productive Well," see **Chapter Eight**. As to comparable problems under Dry Hole Contribution letters, see **§ 2.04(d)**.

It's been suggested that the problem can be solved by using "commercial quantities" (profitable well) or "paying quantities" (Habendum Clause well). While some courts have drawn this distinction, others have held that the terms are synonymous. See *Texaco, Inc. v. Fox*, 228 Kan. 589, 618 P.2d 847 (1980).

[82] As to such subsequent joint operations, see **Chapter Four**.

Problems and pitfalls in defining the Farmee's testing obligations are discussed in § **2.04(b)** and its subsections. The Farmor, however, may reasonably be expected to impose more stringent testing conditions under a Farmout arrangement than might be looked at as reasonable in a Support Agreement.

The Farmor may also impose certain additional requirements as a condition of earning under a Farmout arrangement, particularly where the acreage is exploratory in nature. Among the more common of such provisions would be to impose a seismic obligation on the Farmee.[83] Although it would be more unusual, the Farmor may also wish to provide additional conditions to the earning such as the use of a affiliated company of the Farmor as the drilling contractor for the earning wells, or an agreement on the part of the Farmee to grant a Call on Production not related to the Farmout Area.[84]

§ 3.04 Interests Assigned and Reserved

§ 3.04(a) In General. As mentioned in § **3.01(b)[4]**, there was little negotiation prior to 1977 concerning the nature of the interest to be assigned to an earning Farmee under the Farmout, and the interest to be reserved by the Farmor. To permit the Farmee to deduct 100% of his Intangible Drilling and Development Costs (see § **1.03(b)[6]**), the Farmee invariably would be assigned 100% of the Working Interest in any earning well *Drilling and Spacing Units* until Payout. The Farmor would reserve an Overriding Royalty Interest during this Payout period, which, at the Farmor's option, could be

[83] The drafting of such an obligation is analogous to similar provisions in a Seismic Option. See **Chapter Five**.

[84] The Farmee, on the other hand, as a condition of accepting the Farmout, may wish to require that *his* rigs be used in drilling the earning wells and for subsequent drilling on the Farmout Area, or that *he* receive a "call" on the Farmor's share of production.

As to Calls on Production, see § **3.04(c)**.

converted to an undivided Working Interest after Payout. As to any *Outside* or *excess* acreage — acreage earned by the Farmee lying outside the earning well Drilling and Spacing Units — the Farmee would normally be assigned an undivided Working Interest throughout this excess acreage, with the Farmor reserving an undivided Working Interest.[85] Under the conventional, pre-1977 Farmout, the Farmout would be *Assigned when Earned* — the acreage would not be assigned to the Farmee until after the earning wells had been drilled and the other earning conditions satisfied.

The size of the Overriding Royalty Interest and the convertible Working Interest to be reserved by the Farmor, and the undivided Working Interest in the Outside Acreage to be earned by the Farmee, were subjects for negotiation. The basic structure of the pre-1977 Farmout, however, was almost universally fixed along the lines just indicated.

As we will see in § **3.04(b)[2]** and its subsections, recent tax developments have radically changed the basic format of many Farmout deals. These changes affect not only the interests in production and acreage to be assigned and reserved, but also the *timing* of the assignment.

Originally, tax considerations precipitated the possible change in Farmout structure. However, this re-thinking of structuring the deal originally motivated by tax reasons, also served to break the long-established pattern of *always* structuring Farmout arrangements along the "traditional" lines, without considering whether other solutions were available which might prove more responsive to the needs of both parties to the deal. (See § **3.04(b)[4]** *et seq.*)

Today's Farmout arrangements are far more varied, and often much more responsive to the objectives of the parties, than the pre-1977 Farmouts.

---

[85] Under a less common variation, acreage would be reserved in a "checkerboard" pattern. See § **3.04(b)[3][c]**.

The earning of Outside Acreage is discussed in § **3.03(b)[6]** *et seq.*

In addition to questions of what interest in production and acreage should be earned by the Farmee, and what interest reserved by the Farmor, Farmout negotiations concerning interests assigned and reserved must also consider certain *Operational Reservations* which the Farmor should make. (See § **3.04(c)**. Certain of these reservations, such as various *Take-Over Rights*, are "standard" in almost all Farmout Agreements. Others, however, such as *Calls on Production* and reservations of *Gas Processing Rights*, are subjects for negotiation from deal to deal.

§ 3.04(b) Interests to be Assigned and Reserved in Production and Acreage.

§ 3.04(b)[1] In General. As mentioned in the preceding subsection, interests to be assigned and reserved in production and acreage were, prior to 1977, generally fixed by tradition. This was based on a perceived impact of the Federal income tax laws on the deductibility of the Farmee's Intangible Drilling and Development Costs. This "traditional" structuring, however, has been greatly affected by an Internal Revenue Service interpretive ruling — designated *Revenue Ruling 77-176.*

We will see below that Revenue Ruling 77-176 has lead informed Farmors and Farmees to reconsider how their Farmout arrangements should be structured to avoid potentially disastrous tax consequences which may arise out of the "traditional" Farmout structure. The Revenue Ruling has also opened the eyes of many explorationists to more effective ways to put together Farmout arrangements.

Unfortunately, a number of oil companies — both Independents and majors — are still taking an ostrich-like approach to the very real impact of the Revenue Ruling. Their position is that the people actually negotiating exploration contracts need not worry about tax matters: taxes are the concern of the tax department![86]

Because of its potentially catastrophic effect, you, in negotiating your contracts, simply cannot ignore the effects of Revenue Ruling 77-176. It is too late, after the "basics" of a Farmout have been negotiated, to go back and totally re-do the deal. The *basics* of the deal structure are greatly affected by this Revenue Ruling. Therefore, it is imperative that you become familiar with exactly *what* the Revenue Ruling would to to traditional Farmout arrangements, and how to structure your Farmout arrangement to take these consequences into account. And, since one impact of the Ruling has been to open avenues to more responsive structuring of Farmouts, an awareness of your alternatives will permit you to structure a better deal for your company (and possibly for the other party to the deal as well!).

§ 3.04(b)[2] Impact of Revenue Ruling 77-176 and Related Tax Problems on the Structure of Farmout Agreements. In General.

§ 3.04(b)[2][a] Taxation of an Exchange of Property for Services. To understand the impact of Revenue Ruling 77-176, you must consider the consequences under the tax laws of an *Exchange of Property for Services.*

As a general rule, if I receive compensation from you for performing services, that compensation is treated as Ordinary Income, taxed to me at Ordinary Income tax rates. (See § **1.03(b)[2]**.) While this creates few special tax problems when I'm paid in *money,* a valuation problem arises if my "compensation" is in the form of *property.* Thus, if you pay me $50 for mowing your lawn, I have received $50 in Ordinary Income. However, if you *give me a lawnmower,* what **amount** of "compensation" have I received, for Federal income tax purposes, for the services?

[86] As to the necessity for *all* explorationists to be aware of the impact of the Federal income tax laws in structuring their deals, see the **Introduction** and § **1.03(b)[1]**.

Under our Federal income tax laws, a person who has received property in exchange for services is treated as having received Ordinary Income equal to the Fair Market Value of the property received, as of the date it was assigned to him. So, in the foregoing example, if the Fair Market Value of the lawnmower I received was $75 as of the date of that "gift," I would be treated, for tax purposes, as having received $75 in Ordinary Income.

Let's look at the opposite side of the Exchange of Property for Services coin. If you were to sell me a piece of property, you would pay tax on your *taxable gain* — the difference between the *amount realized on the sale* and your *Adjusted Basis* in the property. So, if you sold me a lawnmower for $50, and your Adjusted Basis in that lawnmower was $30,[87] you would have a taxable gain of $20. If you had held that lawnmower for over 12 months, your gain could be reported at Long Term Capital Gain rates, assuming that the lawnmower was not an *Inventory Asset* held for sale to customers in the ordinary course of your business. (See § **1.03(b)[2]**.)

But what if, instead of paying for the lawnmower in cash, I "purchased" the lawnmower by mowing your lawn? Here, again, we are faced with the problem of *valuation.*

Logic suggests that, if in determining the Fair Market Value of property received for "compensation for services" purposes, we look at the Fair Market Value of that property, then, in determining the Fair Market Value of services received, we should also look at the Fair Market Value of the services as of the date they were performed. This being "logical," it is, of course, *not* the basis used for Federal income tax purposes.[88] It is considered too difficult to value services for such a computation. Instead, it is presumed that if the Exchange of Property for Services represented an arms-length

---

[87] This Adjusted Basis would be arrived at by adjusting your initial *Basis* — the original capitalized cost of the property — to reduce it by any subsequent amortization deductions, such as Depreciation, and to increase it by any additional non-deductible expenditures.

[88] As to the arbitrary nature of many provisions of the Internal Revenue Code, see § **1.03(b)[1]**.

transaction between two unaffiliated parties, you would never have given up the property for the services unless they were of equal value. So, for purposes of valuing what you have received in exchange for your property, it is assumed that the services received were equal in value to the Fair Market Value of the property given in return for those services — again, as of the date of the assignment of the property. Thus, if the Fair Market Value of your lawnmower, as of the date you "gave" it to me, was $75, you would be treated, for Federal income tax purposes, as having received $75 on the "sale" of your lawnmower. Your taxable gain would then be calculated by deducting your Adjusted Basis in the lawnmower from this $75 Fair Market Value.

Three things should immediately be apparent concerning the taxation of an Exchange of Property for Services:

1. *The Value of the Property Exchanged* will be a determining factor in arriving at the tax consequences of the exchange to *both* parties to the transaction;

2. Considerable *Uncertainty and Disagreement* may arise in valuing that property; and

3. *No Money* will have *Changed Hands* during the course of the transaction: both parties will incur *"Phantom Income"* — a tax liability on "paper profits," when no dollars have actually been received with which the tax can be paid — as a result of

the transaction.[89]

§ 3.04(b)[2][b] Application to Oil and Gas Drilling Situations. The Exchange of Property for Services principles apply to any such exchange, and not merely to oil and gas transactions. Let's assume, though, that I wish to acquire an interest in certain leases owned by you. I approach you for a Farmout; however, you are not interested in seeing a well drilled on these leases. Instead, you wish certain other, non-adjacent leases tested — leases in which I have no interest.

A possible way to accomplish both of our objectives is an *Off-Lease Farmout* — an arrangement under which, by drilling on the acreage which *you* wish to see tested, I acquire an interest in the acreage in which I am interested. However, despite the obvious advantages to both of us from such an arrangement, the Off-Lease Farmout has not, in the past, seen much use in the oil and gas industry, primarily due to the unfavorable manner in which such as transaction is treated for Federal income tax purposes.[90]

The Off-Lease Farmout is treated as an Exchange of Property for Services. In other words, I am performing services (the drilling of a well) *in return for an interest in property* (the assignment of an

89 *"Phantom Income"* — the taxation of "paper profits" before they are actually received (or where no true profit actually exists) — creates triple problems. First, even if a true profit in fact exists, to be realized in cash at some future date, a *premature taxation* has been triggered. This is damaging for Time Value reasons. Second, there may be no true profit: a breakeven transaction or an actual economic loss may have occurred. Here, the taxation of *illusory "taxable profit"* adds damage to an already disappointing project by creating or increasing the economic loss through the taxes payable on the artificial profit. Finally, in either event, there is *no cash on hand with which to pay the tax,* either because the cash will not be received until some future date (the "premature taxation" situation) or because no profit will ever occur (the "illusory profit" case). Whatever the reason, the taxpayer will have to dip into his own pocket — not merely divert cash from the transaction — to pay the tax.

As to Phantom Income in general, and the need to avoid it in structuring oil and gas deals, see **§ § 1.03(b)[1]** and **1.03(b)[8]**. As to Time Value of Money, both in general and as it applies to taxes, see **§ § 1.03(b)[2][c]**, **1.03(b)[8]**, and **1.03(c)[2]** and its subsections.

90 For a discussion of Off-Lease Farmouts, see **§ 3.03(b)[2]**.

interest in certain leases.) As a result, I am treated as having received Ordinary Income equal to the Fair Market Value of the interest in the leases assigned to me as of the date of the assignment. You are treated as having "sold" those leases to me for an amount equal to their Fair Market Value as of the date of their assignment.[91]

It should be readily apparent that several of the negative consequences of an Exchange of Property for Services outlined in § **3.04(b)[2][a]** are present in the Off-Lease Farmout situation. First, it may be quite difficult to establish the Fair Market Value of these undrilled leases as of the date they were assigned to me. (You can be sure that the IRS will take the position that the value of the leases was quite high), since this will increase the taxes payable.)

Second, since no cash has been involved in the transaction, both of us will be required to pay a *"Phantom Income"* tax on dollars we've never seen (and may never see); and at a time when neither of us will have the money available to pay that tax, except out of our other financial resources.

A third consequence is that we have swung 180 degrees away from a primary objective in the tax structuring of our oil and gas transactions, i.e., to *defer* the tax to the latest point in time, and have instead *accelerated* the date on which a tax liability will fall due, thus creating a premature tax liability, in contravention of all Time Value principles.

But if these are the tax consequences of the Off-Lease Farmout, won't the same tax results flow under a traditional Farmout arrangement? Again, it would appear that for tax purposes there will be an Exchange of Services (drilling) for an Interest in Property (the Farmout Leases).

If the customary Farmout was taxed in the same manner as conventional Off-Lease Farmouts, it is doubtful that the Farmout would have enjoyed its historical (and historic) popularity as a key

---

91 These undesirable tax consequences of the Off-Lease Farmout can be avoided if the deal is structured properly: see § **3.03(b)[2][c-5-b-4]**.

means of processing acreage, since the consequences of having a contractual arrangement treated as an Exchange of Property for Services is that the economics of the deal to *both* parties will be adversely affected.

The Farmout arrangement must be neither so "sweet," after adjustment for risk, that the Farmor is unwilling to assign the acreage in return for drilling, nor so "lean" that the Farmee is unwilling to accept the Farmout. Thus, to adversely affect economics by requiring the premature payment of a Phantom Income tax, usually destroys the economics for both parties to the deal. Therefore, neither party will be in a position to "give" — to make the deal more advantageous to the other party so that he can afford to accept it.

Fortunately for the oil and gas industry, it originally appeared that the Internal Revenue Service would not treat a traditional Farmout as an Exchange of Property for Services. In a 1941 interpretive ruling — *G.C.M. 22,730* — the IRS announced that, in the traditional Farmout, the Farmor and Farmee were "pooling their capital" for the conduct of drilling activities, rather than engaging in a Sale or an Exchange. As a result, the transaction was treated as a *Sharing Arrangement,* which creates no present tax liability for the Farmor or the Farmee.[91]

As a result of G.C.M. 22,730, until 1975 Farmout arrangements were used extensively by the industry without any fear that adverse tax implications might arise merely as a result of entering into the Farmout arrangement. In 1975, however, the IRS for the first time informally indicated that it had serious questions concerning traditional concepts on the taxation of Farmout transactions when applied to the "broad area" Farmout — where the Farmee was acquiring less than 100% of the Working Interest in the Outside Acreage. The Internal Revenue Service instead indicated that it was severely questioning whether the Farmee should be entitled to deduct 100% of his Intangibles just because he owned 100% of the

---

[91] As to Sharing Arrangements in general, see § **1.03(b)[7]**.

pre-payout Working Interest in the *earning well DSUs alone.* (See § § **1.03(b)[6]**, **3.01(b)[4]**, and **3.04(a)**.)

The proposed change in the perceived former position of the Internal Revenue Service led to a tremendous outcry on the part of the industry, with numerous formal and informal objections wafting their way to Washington. When, in 1977, the IRS formalized its position in *Revenue Ruling 77-176*, the result was a perfect example of "Be Careful What You Ask For — You May Get It!"

The Internal Revenue Service in Revenue Ruling 77-176 did not take away the deductibility of the Farmee's Intangibles (assuming that the traditional tests were met). Instead, the IRS did something much worse!

The fact situation involved in Revenue Ruling 77-176 is illustrated in **Figure 21**. At the time of the Farmout, the Farmor owned 100% of the Working Interest throughout a single undeveloped tract of land. In accordance with the traditional Farmout arrangement, the Farmor agreed to assign the Farmee 100% of the Working Interest in the earning-well Drilling and Spacing Unit, subject to the reservation of a convertible Overriding Royalty Interest; and further agreed to assign the Farmee an undivided one-half Working Interest in the acreage outside the earning-well Drilling and Spacing Unit (the *Outside Acreage*), the Farmor reserving the remaining undivided one-half Working Interest in the Outside Acreage. The acreage was to be assigned to the Farmee after the earning well was drilled.

The holding in Revenue Ruling 77-176 turned upon the impact of an assignment (and reservation) of *differing interests* in what was formerly a single "Property" for Federal income tax purposes. Since upon assignment the Farmee would receive (and the Farmor would now own) separate and differing economic interests in the earning-well Drilling and Spacing Unit and the Outside Acreage, the Outside Acreage and the earning-well DSU were now each separate *"Properties"* for Federal income tax purposes.[92]

[92] As to the "Property" concept, see § **1.03(b)[3]**.

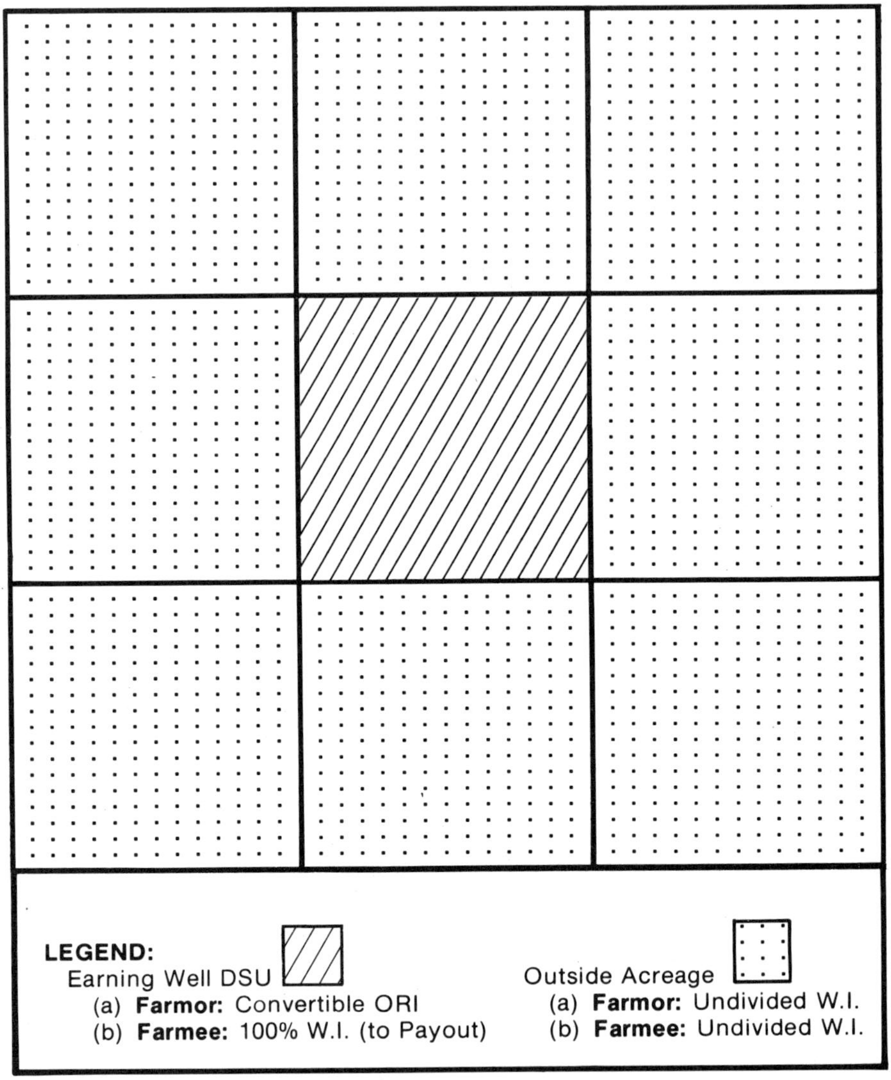

Figure 21: Facts Involved in Revenue Ruling 77-176

The result:

1. **As to the earning-well Drilling and Spacing Unit acreage** (since the earning well was being drilled on the same "Property" which was to be assigned), *the transaction would be treated* as falling within G.C.M. 22,730, i.e., as a *Sharing Arrangement;* and **neither the Farmor nor the Farmee** would *realize income* as a result of the Farmout;

2. **As to the Outside Acreage, the transaction would be treated as an Off-Lease Farmout**, since the acreage being assigned was not the same "Property" which was being developed through the drilling.

Accordingly:

a. **The Farmor would be treated as having received Ordinary Income equal to the Fair Market Value of the undivided interest being assigned to him in the Outside Acreage.** This would be determined as of the date on which the acreage was actually assigned to the Farmee, or on which the Farmee was entitled to an assignment, whichever is earlier.

b. **The Farmor would be treated as having sold an undivided interest in the Outside Acreage for an amount equal to its Fair Market Value** as of the same date. However, the Farmor would be entitled to increase his Adjusted Basis in his reserved Overriding Royalty Interest by the Fair Market Value of the Outside Acreage assigned as of such date.

A copy of Revenue Ruling 77-176 is contained in **Volume Five.**

It's possible that the only "favor" the IRS did the industry with Revenue Ruling 77-176 was that it does not have retroactive application. Thus, the Revenue Ruling applies only to Farmout Agreements entered into prior to April 27, 1977.

Your first reaction may be to think, "So what? Just how serious is this Ruling's impact?" The answer to this question may be "Very," "Ho-Hum," or somewhere in between, depending on the facts and circumstances involved in the particular Farmout transaction.

§ 3.04(b)[2][c] Evaluating and Solving the Tax Problems

§ 3.04(b)[2][c-1] In General. The practical significance of the impact of Revenue Ruling 77-176 on the Farmor and the Farmee varies, depending upon the conditions surrounding the Farmout. In some circumstances, the Revenue Ruling will simply *not be applicable,* because only one property is involved in the transaction. In other situations, a *legal* issue may exist — i.e., the Reveune Ruling will be *legally applicable* — but the *economic impact* of the Farmout may be relatively insignificant to both parties.

Sometimes, though, not only will the Revenue Ruling *legally* apply to the Farmout, but the *economic implications* to the parties will be so *potentially catastrophic* that some form of relief will be absolutely necessary to keep the deal from foundering (or the parties from foundering if the deal goes through as proposed!)

We will discuss various solutions for dealing with the problems created by Revenue Ruling 77-176 in this section. Each of these solutions has certain adverse implications of its own. To trigger potentially disastrous consequences by simply ignoring the Revenue Ruling makes no sense. Accepting "partial" solutions which, though perfectly adequate to deal with certain Farmouts, do not solve the problems when applied to others (including *this* deal!), is also a short-sighted "answer" to the problem.

On the other hand, it is not in your company's best interest for you to unnecessarily create problems where: (i) the Revenue Ruling would not be applicable; (ii) it would have little or no practical economic impact; or (iii) could be adequately solved by less complicated remedies.

Solutions to the Revenue Ruling problem must be considered at the time the Farmout Agreement is being negotiated. It is too late, after the deal has been structured (or even worse, acted upon), to ask your tax department if a potential tax disaster has been created.[93]

It's important to remember that the economic impact of Revenue Ruling 77-176 is usually much more serious on the Farmee than it is on the Farmor. The Farmor is taxed on the *full amount* of the Fair Market Value of the assigned interest in the Outside Acreage, and not just on the *gain* element. The Farmee will also be taxed at the higher *Ordinary Income* tax rate, rather than at the greatly-reduced rate that often applies to a Farmor who can claim a Long Term Capital Gain.

The economic impact of a tax on Phantom Income will also vary, depending on whether the taxed party is a major or an Independent. A tax on Phantom Income will be bad for *any* oil company, including one listed on the New York Stock Exchange. However, since Independents traditionally commit most (if not all) of their present cash flow and cash assets, and their full borrowing power, to present operations, the imposition of the additional cash drain required to pay a tax on Phantom Income could lead to financial ruin. Likewise, since an Independent is usually financing at least a part (and often most) of his participation in the deal through dollars provided by so-called "Tax-Oriented Investors," what to a major might be a "minor" tax negative could prove of real significance to an Independent.

(If the Independent's investors find that $100,000 in Intangibles have been partially or more than offset by large amounts of Phantom Income, they, at best, will turn to other avenues for investing their risk capital. And if investors dry up for the Independents, Farmees tend to dry up for the majors!)

Your job, in analyzing the impact of Revenue Ruling 77-176 on a given trade, is to answer the following questions:

---

[93] Frequently, your tax or accounting department will be dealing with the Revenue Ruling problem in a theoretical "vacuum." However, it is impossible to determine whether or not the Revenue Ruling does or does not create *serious economic problems for your company* (whether you're the Farmor or the Farmee) until you have run a "Quick and Dirty" analysis of the Ruling's economic impact as applied to this particular Farmout. See § **3.04(b)[2][d]**.

(1) **Does the Revenue Ruling even apply?** Many companies have gone through complicated gyrations "to avoid the impact of the Ruling" when the Ruling didn't even apply: legally, there was no Exchange of Property for Services. (See **§ 3.04(b)[2][c-2]**.)[94]

(2) **If the Ruling does apply, how serious is its PRACTICAL impact?** The problem may be so insignificant in its economic impact that no solution is called for, even though the Ruling, from a technical standpoint, is applicable.[95]

(3) **If the Revenue Ruling will have a serious impact, what is the MOST APPROPRIATE SOLUTION under the circumstances?** Most solutions to the Exchange of Property for Services issue create complications of their own. The complications created, and the degree to which the solution will solve the problem — both practically and legally — vary from solution to solution.

You don't shoot off your own arm to kill a mosquito that has just landed there. But if the "trespasser" is a cobra, a shotgun blast might be an appropriate remedy — if no other "solution" was available, and even if your arm was in the line of fire. So, in choosing the appropriate solution to Exchange of Property for Services problems, you'll have to determine if you're dealing with a mosquito or a cobra! (See **§ 3.04(b)[2][d]**.)

§ 3.04(b)[2][c-2] Situations Under Which the Revenue Ruling is Inapplicable: Farmouts Involving No "Separate Properties." There are some circumstances under which Revenue Ruling 77-176

---

94 However, even if the Revenue Ruling doesn't apply, there still may be other *related tax problems* which do need to be dealt with. See **§ 3.04(b)[2][c-4-b]**.

95 Of course, if there is a readily-available, easy-to-implement solution that won't further complicate the deal, we'll still use it. And some of the "solutions" to Revenue Ruling 77-176 also solve other tax problems that exist independent of *Exchange of Property for Services* problems. These solutions may be used on these unrelated tax issues even when the Ruling is not legally applicable, or creates no practical problems.

will simply have no legal application. For instance, let's assume that a Farmor owns certain leases acquired several years ago at a cost of $25 per acre. The Farmor agrees to assign to the Farmee 100% of the Working Interest until Payout in a single 640-acre Drilling and Spacing Unit if the Farmee will drill a specified earning well. The assignment is not to be made to the Farmee until after the earning well is drilled.

The earning well is drilled and successfully completed. As a result, "holdout" offsetting acreage is now leasing for $200 an acre and more.

Despite the increase in the Fair Market Value of the surrounding acreage, the Farmout has involved a *single Drilling and Spacing Unit.* The cornerstone of Revenue Ruling 71-176 is that, following the assignment of the Farmout Leases, the Farmor and the Farmee will now own *differing interests in separate portions of the Farmout Area.* But the single Drilling and Spacing Unit Farmout does not involve "Separate Properties"; and the Revenue Ruling is simply not applicable.

(We're assuming that the single DSU was the entire acreage involved in the transaction. If the Farmee has an "option" to earn additional DSUs, as in a "continuous drilling" situation — see **§ 3.03(b)[6][b]** — the answer might be different, even though each DSU was covered by a separate Farmout or Drilling Option "Agreement.")

An analogous situation arises where the Farmor assigns and reserves a *Uniform Interest* throughout the entire Farmout Area. For instance, assume that the Farmor owns a solid block of leases — purchased a number of months ago at $25 per acre — covering nine 640-acre Drilling and Spacing Units. The Farmor agrees that, upon the drilling of an appropriate earning well, he will assign the Farmee 100% of the Working Interest in all nine Drilling and Spacing Units, subject to the reservation of a non-convertible and Overriding Royalty Interest throughout the entire Farmout Area.

Here, again, even if the successful completion of the earning well results in a significant increase in the Fair Market Value of the surrounding acreage, the Revenue Ruling will not create adverse tax implications, since there are *no "Separate Properties" created* as a result of the assignment. Neither the Farmor nor the Farmee owns a different interest under leases in one part of the Farmout Area from that which he owns in the balance of the leases.[96]

A word of caution. "Split" assignment of horizons, or the absence of a solid lease block throughout the Farmout Area, could still create problems for the Farmor or the Farmee, even though a single DSU, or a Uniform Interest throughout the Farmout Area, were involved.[97] And, under certain circumstances, some of the solutions applied to the Revenue Ruling problem may still be used to solve other tax problems created for the parties by the Farmout, even though the Ruling itself has no application to the Farmout.[98]

§ 3.04(b)[2][c-3] Situations in Which the Revenue Ruling, Though Applicable, Has Little Practical Impact: Farmouts Involving Small Amounts of "Outside" Acreage or Low Per-Acre Value. Not all Farmouts trigger the adverse tax consequences arising under Revenue Ruling 77-176. Furthermore, in some instances, the Revenue Ruling will be legally applicable; but its *practical*

---

96 The same rule would apply if the Farmor were to retain a convertible Override which could only be converted *as to the entire Farmout Area.*

If the convertible Override applied to the entire Farmout Area, but was convertible on a DSU-by-DSU basis, the situation is much less clear.

97 However, the Farmor doesn't have to own 100% of the *Working Interest* in the Farmout Area, so long as there are no "gaps" in *some coverage* of the area, and a uniform fraction of the Farmor's interest *under each lease* is assigned to the Farmee.

98 For example, if the Farmor would normally be entitled to claim Percentage Depletion, or was concerned over capitalization of Delay Rental reimbursements to the Farmor, a *Present Sublease and Assignment* might still be called for. (See **§ 3.04(b)[2][c-5-a]**.) Or, if the Farmout Agreement had been structured to provide for "Disproportionate Allocation" of costs and the pre-payout Working Interest, a *Tax Partnership* might still be advisable. (See **§ 3.04(b)[2][c-5-b]** and its subsections.)

consequences will be so slight that there will be little reason to take any protective steps that, in themselves, might create their own adverse consequences to the Farmor or the Farmee which outweigh the disadvantages of leaving well enough alone.

Let's look at several examples.

Remember, it's the Aggregate Fair Market Value of the Interest in the Outside Acreage Assigned to the Farmee, as of the date of that assignment, that sets the tax liability for both parties to the Farmout. Accordingly, if there is a *relatively small amount* of *Outside Acreage* that will be assigned, or if the *Fair Market Value of the acreage will still be slight* as of the assignment date, there will be a tax payable. But *that tax will be relatively small* — not enough to justify some convoluted structuring of the Farmout deal: the kind of structuring that would be more of a pain in the drill bit than can be practically justified by the limited amount of additional dollars to be paid to Uncle Sam if the deal is structured in the conventional way.

Figure 22 illustrates the impact of the Revenue Ruling where only a small amount of Outside Acreage will be assigned to the Farmee. In this example, we've assumed that the Farmor had acquired leases over 12 months ago at an initial cost of $25 per acre. The Farmee will earn 100% of the Working Interest in the 40-acre earning-well DSU, and an undivided 50% Working Interest in an offsetting 40 acres, by drilling the earning well.

As of the date of the Farmout, there had been no increase in the Fair Market Value of the acreage. As a result of the successful completion of the earning well, however, lease costs of acreage surrounding the earning well increased to $200 an acre. It is only upon the completion of the earning well, and the balance of the earning conditions, that the Farmee is to be assigned his interest in the acreage.

Tax consequences? Minimal. The Revenue Ruling *is* applicable; and the transaction, under the tax laws, results in an Exchange of Property for Services. Likewise, it could have been anticipated that

	Assign After Well Completed	Assign When Execute FO
1. Number of Net Outside Acres Assigned	20	20
2. *Times* FMV/acre When Assigned	x200	x25
3. FMV of Outside Acreage at Time of Assignment	4,000	500
4. **Tax Consequences to Farmor:**		
a. Amount Realized on Exchange	4,000	500
b. *Less* Adjusted Basis ($25 x 20)	<500>	<500>
c. Gain on Exchange	3,500	-0-
d. *Times* LTCG rate (Corporate)*	x.28	x.28
e. **TAX PAYABLE**	**$ 980**	**-0-**
5. **Tax Consequences to Farmee:**		
a. Compensation for Services	4,000	500
b. *Times* Ordinary Income tax rate (Individual)**	x.50	x.50
c. **TAX PAYABLE**	**$2,000**	**$250**

* If Farmor was an Individual, tax rate would be a maximum of 20%.

** If Farmee was a Corporation, tax rate would be a maximum of 46%.

Figure 22: Impact of Revenue Ruling 77-176: Small Acreage Involved

lease values might increase if the earning well was productive. (And they did!) Nevertheless, the *small amount of Outside Acreage* results in a relatively small *aggregate* (not per-acre) Fair Market Value even where the acreage is assigned to the Farmee after the earning well is completed and the acreage has significantly increased in value. The Farmor's "gain on sale" is only $3,500, leaving a *tax payable* — at the 28% Corporate Long Term Capital Gains rate — of less than $1,000. The Farmee, who is taxed on the *full* amount of the Fair Market Value of the Outside Acreage and not just on the "gain" element, has larger tax to pay — more than twice that of the Farmor. But here, again, the amount is small — $2,000.

Even if several thousand net acres of Outside Acreage will be earned by the Farmee, the Revenue Ruling will have little practical impact if the per-acre Fair Market Value of that acreage will remain relatively low even after a productive well is completed on the Farmout acreage.

In **Figure 23**, the Farmor owns leases covering nine potential 640-acre Drilling and Spacing Units in a "mature" area. These leases were purchased two years ago at a cost of $2 per acre, and, as of the date of the Farmout, were worth $5 an acre.

The Farmor agrees to assign the Farmee 100% of the Working Interest in the earning well DSU, and an undivided 50% Working Interest in eight direct and diagonal offsets, if the Farmee will drill a specified earning well. The acreage, again, will be assigned to the Farmee only after the earning well is drilled. Due to the mature nature of the area, and despite the "successful" completion of the earning well, lease prices of surrounding acreage rise only to $6 per acre.

Again, Revenue Ruling 77-176 is legally applicable: the Farmor and the Farmee will now own differing interests in the earning well DSU and in the Outside Acreage, thus creating separate "Properties" in these different portions of the Farmout Area. Also, significant Outside Acreage is to be assigned — several thousand net acres. Despite the *legal ramifications* (in other words, there has been an

	Assign After Well Completed	Assign When Execute FO
1. Number of Net Outside Acres Assigned	2,560.00	2,560.00
2. *Times* FMV/acre When Assigned	x6	x5
3. FMV of Outside Acreage at Time of Assignment	15,360.00	12,800.00
4. **Tax Consequences to Farmor:**		
a. Amount Realized on Exchange	15,360.00	12,800.00
b. *Less* Adjusted Basis ($2 x 2,560)	<5,120.00>	<5,120.00>
c. Gain on Exchange	10,240.00	7,680.00
d. *Times* LTCG rate (Corporate)*	x.28	x.28
e. **TAX PAYABLE**	**$ 2,867.20**	**$ 2,150.40**
5. **Tax Consequences to Farmee:**		
a. Compensation for Services	15,360.00	12,800.00
b. *Times* Ordinary Income tax rate (Individual)**	x.50	x.50
c. **TAX PAYABLE**	**$ 7,680.00**	**$ 6,400.00**

* If Farmor was an Individual, tax rate would be a maximum of 20%.

** If Farmee was a Corporation, tax rate would be a maximum of 46%.

Figure 23: Impact of Revenue Ruling 77-176: Low Per-Acre Value After Well Completion

Exchange of Property for Services), the adverse tax consequences, when reduced to "dollars and cents" terms, are rather minimal. The Farmor has a gain of slightly over $10,000, resulting in a tax, at the corporate Long Term Capital Gains Rate, of less than $3,000. The Farmee's Taxes Payable are greater — nearly $8,000 — but, again, probably not so great as to "blow" the Farmout arrangement.

Looking at the "bottom line," the "seriousness" of Revenue Ruling 77-176, from a practical standpoint, depends on the *aggregate Fair Market Value of the net Outside Acreage being assigned to the Farmee as of the date of the assignment.* This is both a function of the net acres to be assigned and the per-acre Fair Market Value at the time of that assignment. So, if a small amount of acreage is involved, or the per-acre Fair Market Value of that acreage as of the date of the assignment will remain relatively low, tax academics should not get in the way of practicality: there is no need to unnecessarily complicate the deal (or, even worse, adopt a Farmout structure which will create its own unpleasantries for the Farmor or the Farmee), when there will be only a limited amount of additional dollars payable to Uncle Sam despite the *Recognition of Income* under Revenue Ruling 77-176.

(The parties *may* choose to utilize a less conventional Farmout structure for other reasons — tax or otherwise. See § § **3.04(b)[2][c-4-c]** and **3.04(b)[2][d]**. Or, a "fine tuning" modification of the conventional Farmout structure may be adopted to eliminate *any* unnecessary tax, if this can be done without complicating the deal. After all, over numerous Farmouts, those $10,000 to $15,000 per-deal tax liabilities due tend to add up. However, *involved* structures, with serious adverse consequences of their own, should be avoided unless the tax advantages of the more complicated structure are monetarily beneficial.)

One conclusion should now be inescapable: **it is impossible to determine whether or not "drastic" solutions to Revenue Ruling 77-176 are called for unless you "put the pencil" to the Farmout arrangement to find out if the dollar amount of tax con-**

sequences under the conventional structuring (or a proposed "partial" solution) will be minimal or significant.

§ 3.04(b)[2][c-4] Situations Under Which the Revenue Ruling Has Serious Practical Consequences: The Impact of a High Aggregate Outside Acreage Value as of the Date of the Assignment

§ 3.04(b)[2][c-4-a] In General. As illustrated in **§ 3.04(b)[2][c-3]**, the conventional Farmout could create *separate Properties* within the Farmout Area, thus triggering "Exchange of Property for Services" consequences. However, if the *aggregate* Fair Market Value of the Outside Acreage to be assigned will be relatively small *as of the date of the assignment*, no serious practical problems exist under the Revenue Ruling. If the Outside Acreage to be assigned will have a significant aggregate Fair Market Value when assigned, however, the parties must take certain protective steps to avoid adverse consequences under the Ruling.

A number of "Rube Goldberg" solutions to the Revenue Ruling have been developed by "creative" tax strategists. These include "option arrangements" such as assigning the *Farmout Area* a Drilling and Spacing Unit at a time; making separate Farmout Agreements for each DSU within the Farmout Area; assigning only an undivided Working Interest throughout the entire Farmout Area, while assigning the Farmee 100% of the "operating rights" and revenues in the wellbore of the earning well. The solutions may — or may not — have the desired tax consequences.[99]

There are other solutions that might work, *taxwise.* These include retaining a Non-Convertible Overriding Royalty (or Net Profits Interest) throughout the entire Farmout Area, or a Convertible Override which converts only after Payout has been achieved as to the Farmout Area as a whole. These "solutions," however, in most

---

[99] See Larason, "How to Structure the Farmout and Avoid Its Pitfalls," contained in **Volume Five.**

instances won't accomplish the *economic and business objectives* of the parties.[100]

Where the aggregate Fair Market Value of the Outside Acreage to be earned by the Farmee might be relatively high following the completion of the earning well,[101] the industry has primarily turned to two solutions:

(1) The "Present Assignment" (see § **3.04(b)[2][c-5-a]**);

(2) The "Tax Partnership" (see § **3.04(b)[2][c-5-b]** *et seq.*).

The Present Assignment approach is generally viewed as involving virtually no serious practical disadvantages to the parties. (As we'll see in § **3.04(b)[2][c-5-a]**, this is not necessarily the case.) The Tax Partnership admittedly has disadvantages, which may or may not be outweighed by the favorable consequences which result from its use. But, above all, most members of the industry have heard of both solutions; and few understand when or why one, the other, both, or neither should be used. (The same is true of most tax and accounting personnel.)

The result? An indiscriminate use of various solutions, often when they are unnecessary (and even damaging); coupled with an ignorance (or mis-solution) of the problems when "drastic" action *was called for.*

The availability of plentiful drilling capital in the late 1970s, and the shortage of acreage, saw a change in a traditional industry pattern.[102] More and more, the Farmor was able to turn his Farmout

---

100 As to the various interests which a Farmor may assign and reserve, and the economic considerations favoring one over another, see § **3.04(b)[3]** *et seq.*

101 It is impossible to know, pending the actual drilling of the well, exactly what will happen to acreage values thereafter. If there is a *potential* that values in the aggregate — and, therefore, potential tax liabilities — could be quite high, some "solution" to the *Exchange of Property for Services* problem is normally called for.

102 As to the use of Outside Acreage as an incentive (and necessity) in most exploratory Farmouts, see § **3.03(b)[6]** *et seq.*

without committing Outside Acreage to the deal. In other words, the Farmee, except in the Rockies, would earn only a DSU at a time.

So, up until the 1982-1983 period, the *practical* implications of Revenue Ruling 77-176 were less serious, since very little Outside Acreage was involved in most Farmouts. (This didn't stop the "solutions," though, since few oilmen were analyzing the dollars-and-cents impact of the Ruling on their deals, or understood when the Ruling did — and did not — legally apply.) However, as the recession of the 1970s caught up with the oil industry in 1982, available drilling capital was reduced for majors and Independents alike; and deal terms started to return to normal. (And the likelihood of serious consequences under the Revenue Ruling began to skyrocket.) Continued indiscriminate treatment (or ignoring) of the problems created by the Ruling thus became unacceptable to anyone who understood the serious threat posed by the Ruling when significant amounts of valuable — or potentially valuable — Outside Acreage was being committed to the Farmout.

So — let's try to unravel the mysteries!

§ 3.04(b)[2][c-4-b] Related Tax Problems — Percentage Depletion, Delay Rentals, IDC Deductions. The conventional Farmout structure, in addition to creating Recognition of Income problems under Revenue Ruling 77-176, created other tax problems for the Farmee. If the Farmee was an Independent Producer, he would normally be entitled to claim Percentage Depletion, unless the acreage acquired by him was "proven" at the time of its acquisition. However, if the acreage was not to be assigned to the Farmee until after the earning well was drilled, would not the acreage, *as of the date of the acquisition,* be "proven," albeit by the Farmee's own drilling? Likewise, since a Farmee is required to capitalize Delay Rentals "paid" by him prior to his *acquisition* of the related leases, (whether paid directly by him or by reimbursement to the Farmor), didn't the *Assign When Earned* approach result in the capitalization of those Delay Rentals paid by the Farmee during the earning period?[103]

The tax problems outlined above may or may not have real legal or practical significance to the Farmee. The Farmee may have "earned" his acreage, at least for Percentage Depletion purposes, when the Farmout Agreement was entered into — particularly if a *mandatory* earning well, with no *Productive to Earn* feature, was involved.[104] If the Farmee is not an "Independent Producer" for tax purposes (see § **1.03(b)[4]**), Percentage Depletion is a moot question. And if Delay Rentals are, as is typically the case, only a dollar (or even several dollars) an acre, their capitalization may have little impact on the Farmee's economics.

But in many conventional Farmouts, the *Assign When Earned* structuring could have serious after-tax implications to the Farmee, aside from any Exchange of Property for Services concerns.

A further tax problem was created by the conventional Farmout structure which has nothing to do with the Recognition of Income problems under Revenue Ruling 77-176, or the *timing* of the Assignment. This problem involved "locking in" a certain structure — assigning 100% of the Working Interest until Payout to the Farmee — to prevent the potential capitalization of a part of the Farmee's drilling costs. (Partial IDC capitalization would normally be required if the Farmee's share of such costs were disproportionate to his Pre-Payout Working Interest.)[105]

---

103 Percentage Depletion, and its potential loss under the *Proven Property Exclusion*, and the tax treatment of Delay Rentals, are discussed in § **1.03(b)[4]**. As to the *Assign When Earned* approach, and Farmee reimbursement of earning-period Delay Rentals, see § § **3.01(b)[5][3]**, **3.04(b)[2][c-5-a]**, and **3.05(b)**.

104 As to *mandatory* versus *optional* earning wells, see § **3.03(b)[3]**. *Productive to Earn* provisions are discussed in § **3.03(b)[7]**.

105 As to the rules concerning the deductibility of Intangible Drilling and Development Costs and its impact on Farmout structure, see § **1.03(b)[6]**.

§ 3.04(b)[2][c-4-c] Highlighting the Tax Problems Created by the "Conventional" Farmout Structure. Recognition of Income concerns under Revenue Ruling 77-176 were, by no means, the only tax problems facing an explorationist trying to structure a deal using the "conventional" Farmout structure: they were simply the problems that were getting the worst "press."

The explorationist needed (and still needs) to consider a number of potential tax consequences in structuring his deal. The confusion surrounding the Revenue Ruling clouds the issue.

Let's focus on *all* the issues you need to consider:

(1) *Recognition of Income:* Have you *unnecessarily* created Recognition of Income through an Exchange of Property for Services, as a result of an intentional or unintentional (e.g., Rev. Rul. 77-176) *Off-Lease Farmout*?

(2) *Percentage Depletion:* Will the Farmee *unnecessarily* lose Percentage Depletion?

(3) *Delay Rentals:* Will the Farmee *unnecessarily* be required to capitalize Delay Rentals paid or reimbursed during the earning period?

(4) *IDC Deductions:* Will the Farmee *unnecessarily* be required to capitalize a portion of his legitimate drilling costs? If not, has the "price tag" for securing those deductions been to utilize a deal structure which would otherwise be contrary to the economic objectives of the parties?

(Note the continued emphasis on "unnecessarily." Remember that maximizing tax savings isn't our goal; taxes are simply *one part* of the economic equation. See § **1.03(b)[1]**. Also, it's the *practical impact* of the tax considerations we're concerned with, *not* whether, as a matter of law, the tax ramifications are academically adverse.)

The problems are now highlighted. On to the solutions!

§ 3.04(b)[2][c-5] Solving the Tax Problems

§ 3.04(b)[2][c-5-a] Present Assignment/Sublease. The practice of assigning leases after the earning well has been drilled has an adverse impact on the parties under a Farmout in various ways. First, assuming a successful well, this means that the leases are being assigned at the point in time when their Fair Market Value will normally have skyrocketed. Furthermore, use of the *Assigned When Earned* approach may imperil an "Independent Producer's" ability to claim Percentage Depletion, and require that Delay Rental payment/reimbursements by the Farmee be capitalized. (See § **3.03(b)[2][c-4-c]**.) If the leases were, instead, assigned *at the time the Farmout Agreement was entered into*, the potential loss of Percentage Depletion and the requirement that Delay Rentals be capitalized would be *totally avoided;* and the impact of the Revenue Ruling, though not eliminated, might be *significantly reduced.*

Often, the major increase in the Fair Market Value of Farmout Leases will occur after the earning well has been drilled, particularly in a "rank wildcat" Farmout Area. If the leases could be assigned prior to the time the earning well was drilled, there would often be little increase in value *as of the date of the assignment.*[106] This reduction in Fair Market Value as of the date of the assignment would not *eliminate* the impact of the Revenue Ruling: there is still an Exchange of Property for Services, and some tax would still be payable, at least on the part of the Farmee. However, by reducing the Fair Market Value as of the date of the assignment, the parties would be reducing *the amount of the tax*, often to levels which, from a practical standpoint, would make the "adverse" tax consequences somewhat insignificant.

Figure 24 illustrates the problems created when a Farmout involves Outside Acreage whose value would be considerable if a successful earning well were drilled. Here, we've again assumed that the Farmor acquired leases more than 12 months prior to the date of the Farmout. The initial cost of the acreage was low — $5 per acre.

---

[106] This assumes that *increase* in value is our only problem, which is simply not the case: see below.

	Assign After Well Completed	Assign When Execute FO
1. Number of Net Outside Acres Assigned	2,560	2,560
2. *Times* FMV/acre When Assigned	x500	x10
3. FMV of Outside Acreage at Time of Assignment	1,280,000	25,600
4. **Tax Consequences to Farmor:**		
a. Amount Realized on Exchange	1,280,000	25,600
b. *Less* Adjusted Basis ($5 x 2,560a.)	<12,800>	<12,800>
c. Gain on Exchange	1,267,200	12,800
d. *Times* LTCG rate (Corporate)*	x.28	x.28
e. **TAX PAYABLE**	**$ 354,816**	**$ 3,584**
5. **Tax Consequences to Farmee:**		
a. Compensation for Services	1,280,000	25,600
b. *Times* Ordinary Income tax rate (Individual)**	x.50	x.50
c. **TAX PAYABLE**	**$ 640,000**	**$ 12,800**

* If Farmor was an Individual, tax rate would be a maximum of 20%.

** If Farmee was a Corporation, tax rate would be a maximum of 46%.

Figure 24: Impact of Revenue Ruling 77-176: Low Aggregate Fair Market Value as of Date of Farmout

The Farmor enters into a Farmout Agreement covering nine potential 640-acre Drilling and Spacing Units at a time when the acreage is worth $10 an acre. The Farmor agrees that, if the Farmee will drill a specified earning well, the Farmor will assign the Farmee 100% of the Working Interest in the earning well DSU, and an undivided 50% Working Interest in the eight direct and diagonal offsets, with the assignment to be made after the acreage is earned. The Farmee is also to reimburse the Farmor for 100% of the Delay Rentals paid on the Farmout Leases during the earning period.

The earning well is drilled by the Farmee, which causes lease values of surrounding acreage to climb to $500 an acre. Following the drilling of the earning well, the Farmee is assigned his interest in the Farmout Leases.

We've just stepped into a real quagmire. First, unlike the previous situations, the aggregate Fair Market Value of the net Outside Acreage to be assigned may be quite high following the drilling of the earning well; furthermore, that aggregate Fair Market Value represents a significant increase over the Fair Market Value of the acreage at the time it was acquired by the Farmor. As a result, both the Farmor and the Farmee will be paying taxes at a "six figure" level — almost $355,000 for the Farmor and a staggering $640,000 for the Farmee, or a total of nearly *one million dollars*! Moreover, the Farmee may be required to capitalize the Delay Rental reimbursements. And — assuming he has Independent Producer status — the Farmee may have placed his ability to claim Percentage Depletion in jeopardy.

The *Present Assignment/Sublease* solution to Revenue Ruling 77-176 involves assigning the Farmout Leases to the Farmee *at the time the Farmout Agreement is entered into*, rather than at the time the earning conditions are satisfied. The main drawback to this approach has been that it might result in a "clouding" of title if the Farmee did not, in fact, satisfy the earning conditions. Even if the Farmout Agreement provided for an automatic termination of the assignment in the event the earning conditions were not satisfied, or permitted the Farmor to record a "conclusive" unilateral statement of non-performance, most title examiners would not be happy with the title. Accordingly, the Farmor might encounter difficulties in entering into a subsequent Farmout Agreement with, or in selling the leases to, another party if the earning conditions were not satisfied, but the Farmee failed to formally reassign the acreage. And the Farmee, in the absence of a subsequent, confirming unconditional assignment, would have a clouded title, even if the conditions were met.

There is nothing in our tax laws, however, that says that the assignment of leases, for tax purposes, must be *recorded*, or even *recordable in form*. And there is no requirement that *legal* title pass, so long as the assignment is effective to transfer *equitable* title. Accordingly, a provision *in the Farmout Agreement* that the Farmor "hereby assigns" to the Farmee the specified interests in the Farmout Leases should be sufficient to fix the date of the Farmout Agreement, rather than the date of the drilling of the earning wells and the delivery of an implementing *recordable* assignment, as the appropriate time for valuing the assigned interest in the Farmout Leases for purposes of the Revenue Ruling. It will also insure that any Delay Rental reimbursements by the Farmee to the Farmor, made after the effective date of the Farmout Agreement, should be deductible by the Farmee. And the Farmee will still be able to claim Percentage Depletion as to production from the earning well, assuming that the Farmee does qualify as an Independent Producer and the leases were not "proven" as of the date of the assignment.

If the aggregate Fair Market Value of the Outside Acreage were relatively low at the time of the Farmout, Recognition of Income problems, though not eliminated, would be greatly reduced — often to manageable levels. The Farmee now owes only $12,800 in taxes —a far cry from the prior $640,000. And the Farmor's tax liability has been reduced from $354,816 to under $3,600.

Thus, in the situation illustrated in **Figure 24**, since the Farmout Agreement will be non-recordable in form, the Farmor's title will not be clouded. (To assure good title to the Farmee, however, the

Farmout Agreement should call for the delivery of a formal, *recordable* assignment after the earning conditions are satisfied.)

It is arguable that the "Terminable Assignment" approach might be treated as a sham for tax purposes. A safer approach, from a tax standpoint, would be to structure the transfer as a "Present *Sublease*," rather than as an assignment. Here, the Agreement would state that the Farmor "Hereby Subleases" the Farmout Leases to the Farmee for an "Initial Term" ending as of the date by which the earning well must be commenced, and so long thereafter as operations are conducted under the Farmout Agreement for the drilling, testing and completing of the earning wells. The Agreement should further provide that so long as the earning conditions are satisfied, the assignment and sublease shall continue in effect for so long as any of the Farmout Leases, or extensions or renewals thereof, continue in effect.

The Present Assignment/Sublease does not *eliminate* the impact of Revenue Ruling 77-176; it merely *reduces the taxes payable* as a result of the Revenue Ruling if the earning wells are productive and leases increase in value.

A "Catch 22" is that, if the earning well is *non-productive*, the leases may *go down in value:* by using the Present Assignment and Sublease, the Farmor and the Farmee have unnecessarily incurred a tax liability.

If *Dry Hole* drilling under the Farmout condemns the Farmout Leases, this tax negative of the Present Sublease and Assignment approach will not have any great impact on the Farmor or the Farmee, at least so long as the leases are now released. From the Farmee's standpoint, he will now have an offsetting *Abandonment Loss* equal to the Fair Market Value on which taxes were paid (which becomes his "basis" in the Farmout Leases).[107]

---

[107] See § **1.03(b)[4]**.

The Farmor may actually have received a tax windfall from the arrangement: his taxes were payable on the Fair Market Value *increases* of the leases at *Long Term Capital Gain* rates (assuming the leases had been held for over 12 months); but his Abandonment Loss Deduction will be taken at *Ordinary Income* tax rates on his full basis. Thus the Farmor will have accomplished a Long Term Capital Gains conversion. See **Figure 25**.

Where the Present Assignment/Sublease is being used as a solution to the Recognition of Income problems created by Revenue Ruling 77-176 and the further Percentage Depletion/Delay Rental ptifalls mentioned in § **3.04(b)[2][c-4-b]**, it is critical that such present transfer be properly drafted. Many Farmout Agreements provide that the Farmor "Hereby *Agrees* to Assign and Sublease ..." While such a provision may be sufficient to pass equitable title, it seems a risky practice: it would be far better for the Agreement to provide that the Farmor *"Hereby Assigns and Subleases"* the Farmout Leases. Likewise, due to the *Depth Earned* limitation provisions frequently contained in the Farmout, the Farmor may wish to limit the assignment to depths which bear some relation to the depths drilled in the earning well. (See § **3.03(b)[5]** and its subsections.) Accordingly, a number of Farmout Agreements are now providing that "Farmor hereby assigns the Farmout Leases to Farmee *to the stratigraphic equivalent of the depth* **to be drilled** *by Farmee"* in the earning well. Such an attempt to make a Present Assignment/Sublease to a *presently indeterminate depth* is highly questionable; a far better provision would be to make a Present Assignment/Sublease with the provision that the Assignment/Sublease would terminate following the drilling of the earning well as to all depths below the depths under the Farmout.

A sample "Revenue Ruling 77-176 Farmout Agreement," providing for a Present Assignment and Sublease, is contained in **Volume Four**.

Many people have exaggerated the merits and downplayed the potential disadvantages of the Present Assignment/Sublease.

1. Assumed Long Term Capital Gain on Exchange	$100,000
2. Tax Rate (Corporate)	.28
3. Taxes Payable	28,000
4. Loss Recognized on Abandonment	100,000
5. Ordinary Income Tax Rate Otherwise Payable	.46
6. Tax Savings	46,000
7. Tax Savings on Abandonment	46,000
8. *Less* Taxes Paid Due to Farmout	<28,000>
9. NET TAX ADVANTAGE	$ 18,000

Figure 25: Impact on Farmor of Dry Hole Following Present Assignment/Sublease

The explorationist, understandably, desires to find "simple" solutions to the problems created by the Revenue Ruling. And the *Tax Partnership* approach (see § **3.04(b)[2][c-5-b]** *et seq.*) — the most common alternate solution to the Present Assignment/Sublease — has its admitted disadvantages, particularly for the accounting department.

Obviously, if a substantial increase in acreage Fair Market Value is to be expected if the earning well is productive; if the likelihood of a productive well is quite high; or if the Farmee needs Delay Rental/Percentage Depletion protections, the Present Assignment/ Sublease approach should be utilized.[108] More often than not, however, the earning well will be a dry hole or a marginal producer rather than the "bonanza" which would lead to a substantial increase in the Fair Market Value of the Outside Acreage. This is particularly true in the exploratory Farmout; and it is primarily in the area of exploratory Farmouts where a large increase in acreage Fair Market Value could be expected if the earning well is productive. Furthermore, if the earning well, though dry, reduces the Fair Market Value of the Farmout Leases but does not condemn the acreage, no Abandonment Loss Deduction will be available until the acreage is released or the leases otherwise terminate or are condemned. Thus, if a Tax Partnership is not to be formed, and Percentage Depletion loss and Delay Rental problems are not involved, a "balancing act" is called for, to weigh the potential tax liability if the Present Assignment/Sublease is used unnecessarily (the acreage *didn't* go up in value) with the potential taxes payable if the Assign When Earned approach is used and the earning well is a barn burner.[109]

---

108 The Tax Partnership *does not* solve the Percentage Depletion and Delay Rental problems. See **§ 3.04(b)[2][c-5-b-3]**.

109 In addition, if the "dry" earning well is not drilled and the acreage is not condemned and released until the following Tax Year the parties will have lost the *Time Value* of the taxes paid until the Abandonment Loss deduction can be taken in the following year.

Aside from the potential disadvantages of the Present Assignment/Sublease approach, utilizing a Present Assignment does not always reduce the tax liability of the Farmor and the Farmee to manageable levels. A Present Assignment/Sublease will always reduce the *taxes payable* for both parties, assuming that the earning well is successful and the value of the Outside Acreage skyrockets. In some instances, however, the Present Assignment/Sublease will eliminate the Revenue Ruling as a practical drawback to the *Farmor*, but will not reduce the problems to a workable level for the *Farmee*. Furthermore, there are certain instances — unfortunately, all too common — where the problem is not eliminated, from a practical standpoint, for either the Farmor *or* the Farmee.

The Present Assignment/Sublease approach towards solving Recognition of Income problems under Revenue Ruling 77-176 will "solve" the problem for both the Farmor and the Farmee so long as: (i) limited amounts of acreage are involved; (ii) a small initial bonus was paid for the leases; and (iii) the current value of the leases has not significantly increased as of the date of the Farmout. In any event, a Present Assignment/Sublease may be necessary, even where a Tax Partnership (see § **3.04(b)[2][c-5-b]** *et seq.*) or other solution to the Revenue Ruling problem is attempted, to avoid the potential problems of loss of Percentage Depletion or capitalization of Delay Rentals discussed in § **3.04(b)[2][c-4-b]**. If large amounts of acreage are involved, however, and a *high initial bonus* was paid for the Farmout Leases, the Present Assignment/Sublease approach will solve the Revenue Ruling 77-176 problem *for the Farmor* (so long as there has been no significant increase in the Fair Market Value of the leases prior to the Farmout), *but will not solve the problem for the Farmee.* Moreover, where large amounts of acreage are involved, and *the Fair Market Value of the leases has increased prior to the date of the Farmout,* the Present Assignment/Sublease *will not solve the Revenue Ruling problem for either the Farmor or the Farmee.* Under these circumstances, to avoid a potential disaster as a result of the Ruling, a more complete solution is required.

The Present Assignment and Sublease approach has the advantage of simplicity. It does not involve some of the hassles, administrative costs and pitfalls associated with other solutions such as the Tax Partnership. For this reason, the accounting or other departments in the company which will be most burdened by the more complicated solutions will often argue that a Present Assignment/ Sublease solves the Revenue Ruling 77-176 problems in *all* cases, and that any other solutions are "unnecessary" and should be avoided.

As illustrated in **§ 3.04(b)[2][c-5-a-1]**, it is totally untrue that the Present Assignment/Sublease will *always* solve the Revenue Ruling 77-176 problems. It is easy, by making the simple analysis discussed in **§ 3.04(b)[2][d]**, to determine whether or not the Revenue Ruling will still have a significant tax impact even though a Present Assignment/Sublease has been used. Where the Present Assignment/ Sublease will solve the problem or reduce it to manageable levels, it is not to the advantage of your company to attempt more complicated solutions. On the other hand, where the Present Assignment/ Sublease approach would still leave your company subject to a potential tax catastrophe, you shouldn't be bulldozed (or hoodwink yourself) into relying on inadequate protective steps.

§ 3.04(b)[2][c-5-b] The Tax Partnership

§ 3.04(b)[2][c-5-b-1] The Impact of a High Aggregate Outside Acreage Value as of the Date of the Farmout. In those instances where the aggregate value of the Outside Acreage is still relatively low when the Farmout Agreement is entered into (see above), the Present Assignment/Sublease approach solves Recognition of Income problems for both the Farmor and the Farmee, at least in a practical sense. Assume, though, that the Farmor owns leases acquired by him 18 months previously in an area that was already "hot." Initial lease costs were $500 an acre. However, there has been little increase in these values since the leases were acquired; current Fair Market Value appears to be $510 an acre.

The Farmor agrees to farmout nine sections in which he owns 100% of the Working Interest in return for the Farmee's agreement to drill a deep test. The Farmee will earn 100% of the Working Interest in the earning-well Drilling and Spacing Unit, and an undivided 50% Working Interest (a net 2,560 acres) in the remaining eight sections.

The earning well is successfully completed as a prolific producer. Therefore, certain "holdouts" owning mineral rights adjacent to the Farmout Area are now able to lease their acreage for $1,500 an acre. If the Farmout was of the *Assign When Earned* variety, the tax consequences would be catastrophic *to both the Farmor and the Farmee.* However, if the acreage is assigned to the Farmee at the time of executing the Farmout Agreement — the *Present Assignment* solution — the *taxes payable* by the *Farmor* are reduced from over $700,000 to less than $7,200, since the Farmor pays only on the "gain" element and can thus reduce his *amount realized on Exchange* by his original high Basis in his leases.

Not so for the Farmee! The taxes which he must pay on his *Compensation for Services* is based on the *full aggregate value* of the Outside Acreage as of the date of the assignment — a value which was already substantial due to the high bonus initially paid for the leases. Accordingly, while his *tax payable* has been reduced from nearly $2,000,000 to $652,800, this six-figure tax is not "insignificant."

In situations where a high bonus was paid initially for the Farmout Leases, the mere fact that there has been no *increase* in lease value as of the date of the Farmout will not reduce the *taxes payable* to a manageable level for Farmee. See **Figure 26.**

It's also unrealistic to assume that, in all instances, there will be no increase in acreage value over the price originally paid for the Farmout Leases. Thus, if leases were acquired at a relatively low initial cost, but heightened drilling activity or discoveries in the area have lead to an increase in the Fair Market Value of Farmout Lease acreage prior to the development of *this* acreage, serious tax problems can be created for both the Farmor and the Farmee, which

	Assign After Well Completed	Assign When Execute FO
1. Number of Net Outside Acres Assigned	2,560	2,560
2. *Times* FMV/acre When Assigned	×1,500	×510
3. FMV of Outside Acreage at Time of Assignment	3,840,000	1,305,600
4. **Tax Consequences to Farmor:**		
a. Amount Realized on Exchange	3,840,000	1,305,600
b. *Less* Adjusted Basis ($500 × 2,560)	[1,280,000]	[1,280,000]
c. Gain on Exchange	2,560,000	25,600
d. *Times* LTCG rate (Corporate)*	×.28	×.28
e. **TAX PAYABLE**	**$ 716,800**	**$ 7,168**
5. **Tax Consequences to Farmee:**		
a. Compensation for Services	3,840,000	1,305,600
b. *Times* Ordinary Income tax rate (Individual)**	×.50	×.50
c. **TAX PAYABLE**	**$1,920,000**	**$ 652,800**

* If Farmor was an Individual, tax rate would be a maximum of 20%.

** If Farmee was a Corporation, tax rate would be a maximum of 46%.

Figure 26: Impact of Revenue Ruling 77-176: High Bonus Paid for Leases (No FMV Increase)

a Present Assignment will not necessarily solve.

Figure 27 illustrates the problems that develop when the *Fair Market Value of lease acreage has increased as of the date of the Farmout.* The Farmor initially acquired his leases at a cost of $5 per acre. However, the successful completion of wells within several miles of the Farmout Area has caused the Fair Market Value of adjacent acreage to rise to $500 an acre.

The Farmor again agrees to assign nine sections to the Farmee — 100% of the Working Interest in the earning-well Drilling and Spacing Unit and a 50% Working Interest in the eight direct and diagonally offsetting sections — for the drilling of the required earning well. The earning well is successfully completed, and acreage values further increase to $1,500 an acre.

If acreage were not assigned until after the earning well was completed, both the Farmor and the Farmee would have been faced with a huge tax payable. The Present Assignment reduces this tax liability from seven figures to six figures. However, since the increase in the value of the acreage occurred *prior* to the date of the Farmout Agreement, so that there was both a high aggregate value and an *increase* in the value of the Outside Acreage as of the date of the assignment, *neither the Farmor nor the Farmee is faced with an "insignificant" tax liability.*

The situations just outlined illustrate the problems that arise when Farmout acreage already commands a high aggregate Fair Market Value as of the date of the Farmout Agreement. If this aggregate value is a result of a high initial bonus, and there has been no significant increase in that value as of the date of the Farmout Agreement, a Present Assignment will prove a satisfactory practical solution for the Farmor (who would pay a tax only on the "gain" element), but not for the Farmee. If the acreage has increased in value as of the date of the Farmout Agreement, and there is an adequate combination of significant acreage and high per-acre value, the Present Assignment will not reduce the Recognition of Income problems to a manageable level for either party.

	Assign After Well Completed	Assign When Execute FO
1. Number of Net Outside Acres Assigned	2,560	2,560
2. *Times* FMV/acre When Assigned	x1,500	x500
3. FMV of Outside Acreage at Time of Assignment	3,840,000	1,280,000
4. **Tax Consequences to Farmor:**		
a. Amount Realized on Exchange	3,840,000	1,280,000
b. *Less* Adjusted Basis ($5 x 2,560)	<12,800>	<12,800>
c. Gain on Exchange	3,827,200	1,267,200
d. *Times* LTCG rate (Corporate)*	x.28	x.28
e. **TAX PAYABLE**	**$1,071,616**	**$ 354,816**
5. **Tax Consequences to Farmee:**		
a. Compensation for Services	3,840,000	1,280,000
b. *Times* Ordinary Income tax rate (Individual)**	x.50	x.50
c. **TAX PAYABLE**	**$1,920,000**	**$ 640,000**

* If Farmor was an Individual, tax rate would be a maximum of 20%.

** If Farmee was a Corporation, tax rate would be a maximum of 46%.

Figure 27: Impact of Revenue Ruling 77-176: FMV Increase as of Date of Farmout

Under these circumstances, serious consideration must be given to forming a Tax Partnership.

§ 3.04(b)[2][c-5-b-2] What Is a "Tax Partnership?" *Partnerships* (as defined in Subchapter K of the Internal Revenue Code) are not *taxable entities* under the U.S. Federal income tax laws. (See **§ 1.03(b)[2][b]**.) Therefore, Partnerships (as so defined) pay no tax on their "Partnership" income or gain, and also do not directly claim the tax benefits of any losses, deductions, or credits. Instead, taxable income, gain, loss, deductions, and credits are directly passed through to the "Partners" of the Partnership. However, a Partnership (again, as defined in Subchapter K) is a *reporting entity* which must file an "Information Tax Return" with the Internal Revenue Service.[110]

Thus, for Federal income tax purposes, oil and gas activities conducted by multiple parties will be taxed as follows:

1. If, for tax purposes, the participants are treated as **Co-Owners**, any *deductible expenditures will flow through directly to them* and will be reported by them on their personal tax returns. Similarly, any *revenues will be taxed a single time*, at the Co-Owner level.
2. If, for tax purposes, the activity is treated as a **Partnership**, the Partnership, as mentioned, is not a *taxable entity*. Accordingly, there will be *no tax to pay* on any resulting income or gain *at the Partnership level*, and *the* Partners will pay a single tax on their allocable share of Partnership income and gain. *Deductions and losses will also pass through directly to the Partners*, who will report such

[110] The Revenue Act of 1978 added stiffer penalties for failure to file such returns, and, in some instances, extended the time for audit of the tax returns of the Partners where the Partnership had not filed the necessary tax return.

deductions on their personal income tax returns.[111]

3. If the activity is treated as a **Corporation** for tax purposes,[112] there will be *no "flow through" of the tax benefits* directly to the "participants in the deal — the Corporation *is* a *taxable entity*.[113] Any revenues distributed to the "Shareholders" will be subject to a double taxation — first at the *Corporate*, and then at the *Shareholder* level.[114]

In some instances, the *business advantages* of a Corporate structure — limited liability, centralized control, etc. — will dictate use of the Corporate approach, despite its tax disadvantages. Likewise, the tax disadvantages may be reduced by certain tax concepts such as the "dividends received" credit and the "carry forward" of Net Operating Losses. As a general rule, though, multi-party oil and gas ventures are not structured in a Corporate form unless these business advantages clearly outweigh the tax negatives.[115]

Multi-party oil and gas "joint ventures" will usually be structured, then, for tax purposes, as *Co-Owner* or *Partnership* arrangements. Each has its tax advantages and disadvantages. To appreciate what is involved in electing the Partnership approach, however, it's necessary that you understand exactly what we mean when we say

[111] While both Co-Ownership and Partnership conduct of multi-party activities results in a single taxation of income and gain, and a flow through of losses/deductions/credits both, at the individual participant level, the Tax Return "Schedule" on which the transaction is reflected varies, depending upon whether a Co-Owner or Partnership structure is used.

[112] The "unpleasant" tax consequences, flowing from a Corporate characterization for tax purposes, result whether the venture was intentionally structured as a Corporation, or an "Association Taxable as a Corporation" was inadvertently created.

[113] See § **1.03(b)[2][b]**.

[114] The revenues, when distributed to the participants in the venture, will be treated as "Dividends" — taxable to the *Shareholder* participants, but not deductible to the Corporation.

[115] As to the use of the " 'Subchapter S' Corporation," see § **1.03(b)[2][b]**, n. 10.

that a multi-party arrangement will be considered a *Partnership for Federal income tax purposes.*

A *Partnership*, as defined under Subchapter K of the Internal Revenue Code, encompasses *any joint undertaking where the parties will share in both costs and production*,[116] so long as the undertaking is *not* structured, for tax purposes, as a Corporation, a Trust, or an Estate. So, even though a non-incorporated venture would be characterized as a Co-Tenancy for state law purposes, if costs and production are to be shared, the undertaking will be considered a *Partnership* for Federal income tax purposes — unless the parties elect to be *excluded* from the provisions of Subchapter K.

The Internal Revenue Code does give parties to a non-corporate joint undertaking the option of electing to be excluded from Subchapter K's Partnership provisions.[117] If the parties so elect, they will be taxed as Co-Owners, as outlined above. Absent such an election, though, they will be taxed as *Partners — irrespective of their state law status!*

In the conduct of oil and gas operations, members of the industry have traditionally elected to be excluded from the provisions of Subchapter K rather than to be taxed as Partnerships. The reasons for this can be summarized as follows:

1. A desire to avoid the *disadvantages* of Partnership taxation, such as the requirement for filing the Information Tax Return.
2. A *basic misunderstanding* on the part of most members of the industry as to exactly what a "Tax" Partnership was all about.

The idea of being a "Partner" with another company — particu-

---

116 A joint undertaking, where the parties were to share in *expenses alone*, would not constitute a Partnership under Subchapter K.

117 This really doesn't cost the Government much, since Partners and Co-Owners are taxed in much the same manner, with a single taxation of income and gains and a flow through of deductions, losses, and credits. See above.

larly an Independent — is absolutely abhorrent to most major oil companies. Thus, it is often a misunderstanding of the nature of the *Tax Partnership* that leads most companies automatically to elect to be excluded from the provisions of Subchapter K.

Would, instead of using the term *Partnership* to define Subchapter D activities, that the Internal Revenue Code had called such activities "Artichokes." (Few people would have objected to forming a Tax Artichoke!)

A *Tax Partnership,* then, is not required to be a "legal" Partnership — a entity which qualifies as a Partnership, either general or limited, for *state law purposes.* It is merely a joint undertaking where the parties have elected *NOT to be excluded* from the provision of Subchapter K.[118]

Once the provision electing not to be excluded from the provisions of Subchapter K has been inserted in the Farmout or Joint Operating Agreement, the Tax Partnership has been formed.[119] (Again, this has *not* created any state law *Partnership* liability between the parties.)[120]

The basic effect of the election not be excluded from the provisions of Subchapter K, thus, is merely to change the portion of

---

118 The Internal Revenue Code requires that parties *affirmatively elect* to be excluded from the provisions of Subchapter K to avoid Partnership characterization for tax purposes, if the activity would otherwise be treated as creating a Partnership under Subchapter K. It is probably not necessary, therefore, to formally elect *not* to be excluded from the Partnership provisions of the Code, so long as no election *TO be excluded* is contained in the agreement. However, it is customary industry practice — where the parties desire Tax Partnership treatment — to insert a provision in the agreement containing the election *not* to be excluded from Subchapter K.

119 The mere fact that the Tax Partnership has been formed will not automatically result in all the tax advantages of *Partnership* status — such as the ability to make "Special Allocations" — flowing to the Partners. See § **3.04(b)[2][c-5-b-4]**.

120 Of course, if the agreement has been structured in such a way that a "common law" or "legal" Partnership would have existed in any event, the participants will have Partnership liability. This arises due to the basic nature of the relationship, however, and not because the agreement contains the election not be excluded from the provisions of Subchapter K. (Even if the agreement had contained the Subchapter K exclusion election, this would not have eliminated the common law Partnership liability).

the Internal Revenue Code under which the parties will now be taxed. In other words, instead of being taxed under the general provisions that apply to individual taxpayers, including taxpayers engaged in "Co-Tenancy" activities, the particular activity will now be taxed pursuant to the provisions that apply to *Partnership* activities under Subchapter K.

§ 3.04(b)[2][c-5-b-3] Impact on "Recognition of Income" Under Revenue Ruling 77-176. But what does the formation of a Tax Partnership have to do with our problems under the Revenue Ruling? Quite a bit!

As discussed in § **1.03(b)[7]**, when a party contributes either cash or property to a Partnership in return for an interest in that Partnership, the transaction is not a "taxable event" either to the contributing party or to the Partnership. In other words, there are *no taxes payable* by either the Partner or the Partnership as a result of the contribution. What we have done, then, is to create something akin to a Non-Taxable Exchange or a Sharing Arrangement.[121]

As we saw above, the Present Assignment/Sublease only reduced taxes payable under Revenue Ruling 77-176; it did not eliminate the problem. **The formation of a Tax Partnership totally eliminates any Recognition of Income problems which could have arisen under the Revenue Ruling.** In other words, the Present Assignment/Sublease does not eliminate the problems of the Revenue Ruling; it merely reduces the problem to a level which may, or may not, serve as a practical elimination of the Revenue Ruling as a serious impediment to the proposed Farmout arrangement. The Tax Partnership totally eliminates problems under the Revenue Ruling, since there is no longer an Exchange of Property for Services.

It's important to remember that the Tax Partnership has absolutely no impact on the potential problems of loss of Percentage

[121] As to such arrangements, see § **1.03(b)[7]**.

Depletion or a required capitalization of Delay Rentals by the Farmee. Thus, if these related tax problems — discussed in **§ 3.04(b)[2][c-4-b]** — would create difficulties for the Farmee, it is necessary to couple the formation of the Tax Partnership with a Present Assignment/Sublease.

A sample "Revenue Ruling 77-176 Farmout Agreement" containing both Present Assignment and optional Tax Partnership provisions is contained in **Volume Four**.

§ 3.04(b)[2][c-5-b-4] Other Consequences of the "Tax Partnership" Approach. The elimination of Recognition of Income problems under Revenue Ruling 77-176 by the utilization of the Tax Partnership is, in itself, a major reason to seriously consider the use of this technique in structuring the Farmout Agreement. There are, however, other tax and economic consequences, both positive and negative, that flow from the use of the Tax Partnership. Before an intelligent determination can be made as to whether or not a Tax Partnership should be created, both the positive and negative consequences must be taken into consideration — in other words, it's crucial that you fully understand what can be accomplished by using a Tax Partnership, and the price that must be paid to attain these results.

Under the conventional structuring of a Farmout arrangement as a *Co-Ownership*, if the drilling costs to be borne by a Farmee are greater than his Pre-Payout Working Interest, these "disproportionate IDCs" must be capitalized as Lease Costs. (See **§ 1.03(b)[6]**.) As a result, the conventional Farmout "locked" the Farmor and the Farmee into a structuring arrangement — the assignment of 100% of the Working Interest subject to a convertible Override. (See **§ § 3.01(b)[4]** and **3.04(a)**.) This eliminated from consideration other deal structures, such as various "carried" and disproportionate sharing arrangements. These deal structures, under given circumstances, might be more advantageous from an economic standpoint than the "conventional" structure — if it were not for the disadvantageous tax consequences. See § **3.04(b)[3]** *et seq.* Through use of a Tax Partnership, however, disproportionate sharing — referred to, in tax parlance, as "Special Allocations" — can be used without any loss of IDC deductibility, so long as the Special Allocations have "Substantial Economic Effect."

Under a special provision of the Tax Code, the tax consequences of Partnership activities are allocated to the Partners in the manner provided in the Partnership Agreement, so long as the "Substantial Economic Effect" test can be met.[122] Generally speaking, a provision in the Partnership Agreement which merely allocates to a "Drilling Party" the deductions, losses, or tax credits which are generated by dollars actually expended by him will have Substantial Economic Effect. In other words, the "Substantial Economic Effect" test is satisfied if the allocation of losses corresponds to the manner in which the parties share the economic burdens of the venture.

Just as a Present Assignment/Sublease may be used in a Farmout arrangement where no Recognition of Income problems exist under Revenue Ruling 77-176 to avoid Delay Rentals or Percentage Depletion problems, a Farmout Agreement may contain Tax Partnership provisions to permit Special Allocations even when Revenue Ruling 77-176 would be inapplicable. Utilization of the Tax Partnership for Special Allocation purposes will thus permit "Third-for-a-Quarter" (or "Half-for-a-Whole") arrangements, disproportionate sharing, or Casing Point "carry"/Casing Point Election arrangements.[123] Likewise, even if the "conventional" *Payout* sharing arrangement is desired for economic reasons, use of the Tax Partnership will permit the Farmor to impose limitations on that Payout in the event of unjustified cost overruns or sales of production at "brother-in-law" prices, without a loss of tax benefits to the

---

[122] See IRC, § **704**.

[123] See § **3.04(b)[3]** *et seq.*

Farmee.[124]

The use of the Tax Partnership thus frees the Farmor and the Farmee from their prior "structuring straightjacket," and permits them to consider sharing arrangements which may be more economically advantageous — often to both parties![125]

The Internal Revenue Service has recently issued interpretative rulings indicating that Special Allocations which track the actual charging of the related economic expenditure may not necessarily have Substantial Economic Effect, unless the Partnership Agreement also provides for a "balancing" of the Partners' capital accounts on the termination of the Partnership. While no such requirement seems to be contained in the Tax Code, such a balancing does not require that a Farmor give up his "promotion" on the deal; it only imposes an obligation that any "excess" distributions of revenues to a party, *over and above those permitted under the Farmout Agreement,* or any failure of a party to actually pay for costs which, under the Farmout Agreement, were to be charged to him, be taken into account and corrected when the Partnership is dissolved. So, to avoid any question concerning the Substantial Economic Effect of the Partnership "Special Allocation" provisions, a provision for capital account "balancing" should be added to the Farmout Agreement.[126]

As indicated above, the mere fact that a Tax Partnership has been formed, and that costs are to be shared in a different ratio than production, does not mean that such allocations will automatically be recognized for tax purposes *unless the Partnership Agreement expressly so provides.* Accordingly, if the parties desire to utilize a different cost/production (interests assigned/reserved) arrangement than the "conventional" assignment of 100% of the Working Interest

---

124 See § § **1.03(b)[6]** and **3.04(b)[3][c]**.

125 As to various alternatives concerning the interests assigned and reserved by the Farmor, see **§ 3.04(b)[3][b]**.

126 See the sample "Revenue Ruling 77-176 Farmout Agreement" in **Volume Four**.

until Payout, the following steps should be taken to insure that the Special Allocations will not create adverse tax consequences:

1. **The venture should be structured in such a way that it will be treated as a Partnership for tax purposes.** This can be accomplished with or without the formation of a formal "legal" Partnership, *so long as the arrangement DOES qualify as a Partnership under Subchapter K.* This "Subchapter K" structuring requires:

   (a) **An election** by the parties **not to be excluded from the provisions of Subchapter K;**

   (b) Avoidance of "Association Taxable as a Corporation" pitfalls, usually by providing for *a right to take production "in kind."*[127]

2. **An express provision** should be inserted in the partnership Agreement **providing for the Special Allocation of Partnership gains and losses.**[128]

3. **The "Substantial Economic Effect" tests must be satisfied** by insuring that:

   (a) **Losses are allocated to the party charged with the related expenditure;**

   (b) **Inserting a provision for a balancing of capital accounts** on termination of the Partnership.

If, as is commonly the case, a "tax" rather than a "legal" Partnership is to be formed, the agreement should also contain a provision that, *except for tax purposes,* it is not the intention of the parties to form a mining or other partnership.

The sample "Revenue Ruling 77-176 Farmout Agreement" contained in Volume Four contains suggested language to implement the drafting suggestions outlined above.

---

[127] As to Associations Taxable as Corporations, see § **1.03(b)[2][b]**. As to "Taking Production in kind" under Joint Operating Agreements, see **Chapter Four**.

[128] The Partnership Agreement must expressly provide for the Special Allocations. Otherwise, Partnership gains and losses will be allocated "in accordance with the partner's interest in the partnership...." See IRC, § **704(b)**.

When the Tax Partnership was originally suggested as a method of avoiding Recognition of Income problems under Revenue Ruling 77-176 and/or permitting disproportionate allocation of Intangible Drilling and Development Costs, a number of explorationists expressed concern that the Internal Revenue Service might attack this method of Farmout structuring. It now appears clear, however, that a properly drafted Farmout Agreement utilizing Tax Partnership provisions will be recognized by the Internal Revenue Service.[129]

The advantages of using a Tax Partnership is not limited to Revenue Ruling 77-176 and Special Allocation situations. The Tax Partnership approach may also be used when the parties wish to use an *Off-Lease Farmout.*

As discussed in § **3.03(b)[2]**, in certain instances there are advantages in having the earning well drilled on acreage other than that to be assigned by the Farmor to the Farmee. However, just as the "inadvertent" Off-Lease Farmout created by Revenue Ruling 77-176 created Recognition of Income problems for the parties as an Exchange of Property for Services, the intentional Off-lease Farmout created similar problems. Here, again, the express provisions of Section 721 of the Tax Code permit such an "exchange" without any Recognition of Income, so long as a Tax Partnership arrangement is utilized.

So the advantages of the Tax Partnership are considerable. Unfortunately, as indicated above, there is a "price tag" for securing those advantages — certain disadvantages which are also associated with electing *not* to be excluded from the provisions of Subchapter K.

The key disadvantage of the Tax Partnership arrangement is the cost and administrative hassle of the Information Tax Return which each Partnership — "Tax" or otherwise — must file with the IRS for each year of its operation.

At first blush, the "problems" of filing an Information Tax Return

---

[129] See Private Letter Ruling Number 8133021 (April 30, 1981). *Cf.* Private Letter Ruling 8133028 (May 19, 1981).

for a Tax Partnership may appear to be somewhat insignificant. After all, that's what computers are for!

Unfortunately, filing an Information Tax Return for a Tax Partnership creates a greatly increased workload for the accounting department. An Operator who is maintaining records for "Joint Account" billing purposes knows the costs and *production* which have been charged and credited to the Non-Operators; to file a Partnership "Information Tax Return," however, he must know the Non-Operators' costs *and revenues*, i.e., the price at which that production was sold.

The Non-Operator may consider this confidential information. Furthermore, there will certainly be delays in securing the information.

Where a number of small producers are involved in the Tax Partnership, some of whom may have lax administrative practices, further problems are created. If these more careless or dilatory producers are the Non-Operators, the Operator's accounting department may find it difficult and time-consuming to secure the information necessary to prepare the Tax Partnership Information Tax Return. (And remember, *none* of the participants in the Unit may accurately file their own personal tax returns until they have received the Tax Partnership information on *all* of the Tax Partnerships in which they are participating!). If a lax producer is the Operator, he may neglect to advise his accountant or accounting department of the formation of the Tax Partnership, or his accounting personnel may be dilatory in preparing the return.

Starting to see the problems? Now add just a few more complicating factors: An Independent with "Tax-Sheltered Investors" needs to provide accurate tax information to his "et als." by March 15th, but most majors do not complete their tax returns until September; some of the parties may be utilizing a Calendar Tax Year and others a Fiscal Tax Year; some may utilize Accrual and others Cash Accounting Methods; different tax rules apply to corporations and individuals; different companies may be utilizing different equipment depreciation rules. (The complications become even clearer!)

(And we haven't even discussed the problem of potential penalties under the Revenue Act of 1978 if the Partnership Tax Return is not filed in a timely manner!)

Another complicating factor is that a "premature termination" of a Partnership—"Tax" or otherwise—can create tax liabilities for the venture participants that might outweigh any of the benefits initially gained through the formation of the Tax Partnership. This means that the Partnership cannot be terminated once the earning well has been drilled, but must be kept in effect, with annual Partnership tax filing requirements, preferably until the reserves have been depleted and the leases have been surrendered. Thus, the costs and administrative hassle must continue for a significant period of time. Furthermore, either through neglect or operation of law, the Tax Partnership may prematurely terminate anyway, with potential negative tax consequences to all the parties to the venture.

It is difficult to estimate exactly what level of "in-house" costs may be involved in the formation and continued administration of a Tax Partnership, and what the expense of formation and of continued tax return preparation would be for a company which uses outside accountants for such services. Obviously, the costs will be significantly less in a deal involving few parties. This is particularly true where responsible producers, with responsible accounting departments, see to it that necessary information is promptly furnished the Operator and that the tax returns are promptly filed. A totally different situation will be involved if there are many parties to the deal, particularly where you know, going in, that some are slow in answering any type of communication (not to mention in paying their bills!) or that the Operator tends to overlook "minor" obligations such as filing tax returns. However, as a rule of thumb, "in-house" figures have been estimated at $5,000 for each year during which the Tax Partnership must be kept in effect, or a "present value" cost, at the time of formation, of $40,000 over the life of the Partnership.

Whatever the actual costs may be, certain guidelines now become apparent:

1. *There are advantages and disadvantages to the Tax Partnership approach.* You should not hesitate to form a Tax Partnership where the gains outweigh the negatives. Similarly, Tax Partnerships should not be lightly formed if a minor tax saving is far more than offset by the added costs, hassles, and inter-departmental unpleasantness that is often associated with its creation.

2. *It will be impossible to determine whether the benefits of the Tax Partnership outweigh its negatives without first doing a "Quick and Dirty" study* to see how much money will actually be saved by forming a Tax Partnership. See **§ 3.04(b)[2][d]**.

3. *A major oil company will tend to discourage the formation of Tax Partnerships* unless the tax savings are quite high. *An Independent,* however, who secures his drilling capital from "Tax Sheltered Investors" *may find it imperative to insist that a deal be structured in a Tax Partnership format* — despite the costs over the life of the Partnership, — *to avoid investor tax liabilities which might more than offset the tax deductions secured from drilling the well.* (See **§ 3.04(b)[2][c-1]**.) Under these circumstances, a Farmor may acquiesce in the request for a Tax Partnership, but attempt to negotiate some offsetting payment to cover the increased administrative costs which, from his standpoint, were not required. Similarly, if the Operator is known to be lax in his accounting practices, the Farmor may wish to take over the preparation of the Tax Partnership return, again attempting to negotiate an offsetting Cost Reimbursement.

4. *Don't create unnecessary additional problems for yourself!* If multiple Farmout or Joint Operating Agreements are entered into covering a single well, each containing a Tax Partnership provision, there will be multiple Tax Partnerships

and multiple Tax Partnership Information Tax Returns. Under these circumstances, *forming one "master" Tax Partnership under a "master" Farmout or Joint Operating Agreement may be called for*, even though the agreement may contain separate provisions concerning the rights among various parties to the deal. Also:

(a) *Insure that the Tax Partnership and Special Allocation provisions are properly drafted;*

(b) *Take steps to insure that the Partnership Information Tax Return can, and will, be promptly filed;*

(c) *Don't terminate the Tax Partnership prematurely.*

§ 3.04(b)[2][d] Reaching an Intelligent Solution to Farmout Agreement Tax Problems. We've seen advantages, disadvantages, complications and weaknesses in the two major solutions to the problems created by Revenue Ruling 77-176 and various related tax problems. Now it's time to focus on developing a game plan for an intelligent solution of these tax problems on a deal-by-deal basis.

Step number one is to determine if any LEGAL problems actually exist. If no Separate Properties or Off-Lease Farmout is created by the agreement (see § § **3.04(b)[2][b]** and **3.04(b)[2][c-2]** *et seq.*), and if there are no problems of disproportionate sharing (see § **3.04(b)[2][c-5-b-4]** or capitalization of Delay Rentals or loss of Percentage Depletion (see § **3.04(b)[2][c-4-b]**, rest easy! — you can stick with the conventional Farmout structure. (Of course, now that your mind has been opened to more flexible thinking, you may decide that, for *economic* (rather than tax) reasons, you want to try an alternate structuring approach. And this will not necessarily require that you use a Tax Partnership or some other form of complicated Farmout structure. See § **3.04(b)[3][c]**.)

If there will be Recognition of Income or other tax problems if the conventional Farmout structure is used, move to ***Step Two: Determine how serious, from a PRACTICAL standpoint, are the tax problems?***

(Remember: as a rule of thumb, if the aggregate Fair Market Value of the *Outside Acreage* committed to the Farmout would be relatively small *even if the earning well is productive*, you have little practical Recognition of Income problems.)

You can quantify your answer to this question, however, only by performing a "Quick and Dirty" study to see, from a dollars-and-cents standpoint, just how badly your company (and the other party to the deal) will be hit in the pocketbook if the conventional structure is used.

If the answer to the "practical impact" question is "Very little," then you may either decide to stick with the conventional structure — despite the "academic" tax problems — or, at most, turn to a less complicated solution, such as the Present Assignment/Sublease.[130]

If the practical problems *are* serious, move to ***Step Three: Will the Present Assignment/Sublease, if a legally-applicable solution, reduce the practical problems to manageable levels, or must you consider forming a Tax Partnership?*** Here again, the "Quick and Dirty" study is a must.

(Again, as a rule of thumb, if the aggregate Fair Market Value of the Outside Acreage is relatively small *as of the date of the Farmout* — low initial cost, no subsequent increase in the value of the acreage — the Present Assignment/Sublease will usually resolve Recognition of Income problems from a practical standpoint. If there has been no *increase* in acreage values as of the date of the Farmout but high initial bonuses were required, and there is a significant commitment of Outside Acreage, the Present Assignment/Sublease solves the practical problems for the Farmor, but *not* for the Farmee. If the acreage has significantly increased in value as of the date of the

---

[130] Remember, though, that the Present Assignment/Sublease has problems of its own. See § **3.04(b)[2][c-5-a]**. Also, make sure that *all* tax problems, and not merely Exchange of Property for Services questions, have been taken into account in your equation (see § **3.04(b)[2][c-4-e]**, and that your "solution" is appropriate to the problem: a Present Assignment/Sublease doesn't solve "Disproportionate Sharing" issues, and creating a Tax Partnership does nothing toward resolving questions concerning capitalization of Delay Rentals or loss of Percentage Depletion.

Farmout, and considerable Outside Acreage will be assigned, *neither* the Farmor nor the Farmee's practical Recognition of Income problems are solved by the Present Assignment/Sublease.)

What about this "Quick and Dirty" study? As applied to Recognition of Income problems, you can use the following simple formula:

1. ***Aggregate Acreage Value*** Equals ***Net Outside Acreage*** to be assigned Times ***Fair Market Value*** per acre as of the date of the assignment.

2. ***Taxes Payable (Farmor)*** Equals ***Aggregate Acreage Value*** Less the ***Adjusted Basis*** of the Leases Times ***28%***.[131]

3. ***Taxes Payable (Farmee)*** Equals ***Aggregate Acreage Value*** Times ***50%***.[132]

Similar "Quick and Dirty" studies can be made to determine how much extra taxes would be payable if the Farmee was not able to claim Percentage Depletion, or was required to capitalize a portion of his IDCs or Delay rentals.[133]

How do you determine per-acre Fair Market Value and Adjusted Basis of the Farmout Leases? Remember, we're not trying to win a Nobel Prize. We're only trying to determine the *likely magnitude* of

---

131 Assuming the Farmor is a Corporation and has held the acreage for over 12 months. If the Farmor is an Individual and has held the acreage for over 12 months, you would multiply by 20%. If the acreage has been held by the Farmor for 12 months or less, multiply by 50% to determine the Farmor's taxes payable.

132 The 50% Ordinary Income tax rate is used for "Quick and Dirty" purposes even where the taxpayer is a Corporation and, in fact, would be taxed at a maximum tax rate of 46% on its Ordinary Income.

133 Most companies — including many majors who "avoid" Tax Partnerships in Exchange of Property for Services situations (even when they need them!) — automatically form Tax Partnerships to avoid a loss of IDC deductibility, without considering whether the Tax Partnership cost far outweighs the "saved" IDC deductions. Thus, while in a "Third-for-a-Quarter" arrangement (see **§ 3.04(b)[3][b]**) with *one million dollars* in Farmee drilling costs, use of the Tax Partnership will avoid an extra $125,000 in taxes and is well worth it, use of a Tax Partnership to save $12,500 where only *$100,000* in Farmee IDCs (and a $25,000 capitalization) is involved is highly questionable.

the tax consequences — $200, $200,000 or $2,000,000 — to see if, from a *practical* standpoint, a complicating solution (and *how complicating* a solution) is worthwhile. Thus, for "Quick and Dirty" purposes:

1. *Current purchases* in the area can be looked at to determine the current Fair Market Value. (You'd have to do some crystal ball-gazing and look at results of comparable prospects to determine estimated Fair Market Value following a successful well completion. Since this, however, involves only a question of whether or not, *at a minimum,* to use the less complicated Present Assignment solution, some ballpark estimation will tell you whether potential tax libility looks like only a few pennies or in the high six figures.)
2. *Initial bonus paid* can be used as a measure of the Basis of your leases if the accounting department can't give you the actual *Adjusted Basis* of the Farmout Leases.

Remember, if your company is an Independent, and much of its drilling capital comes from outside investors,[134] a loss of initial deductibility, or an offsetting Recognition of Income, could be viewed as catastrophic by your investors.[135] Thus, $40,000 Farmee *taxes payable* might be enough to cause an investor-funded Independent to request a Tax Partnership, while an informed major might never voluntarily seek the Tax Partnership's complications and extra costs under such circumstances.

(Most majors will accomodate an Independent Farmee under these circumstances, though, to avoid having an otherwise desirable Farmout fall through.)

On the other hand, if your extra taxes payable (or the aggregate taxes payable by both Farmor and Farmee) under a "conventional"

---

[134] Ninety percent of all domestic on-shore wells are drilled, and 75% of all domestic on-shore drilling expenditures are made, by Independents. The vast majority of these activities are financed, at least in part, through outside investors. See § **3.04(b)[2][c-1]**.

[135] Each dollar of Recognition of Income tax liability is the equivalent of a two-dollar loss of initial deductibility to the investor.

structuring are only a few thousand dollars, a Tax Partnership makes little sense; and you should also take a long look at the disadvantages (as well as the advantages) of a Present Assignment/Sublease approach as applied to the particular prospect before automatically incorporating it as a "solution" to your tax problems. See § **3.04(b)[2][c-5-2]**.

To summarize:

1. ***You will have NO PROBLEM under Revenue Ruling 77-176, legal or otherwise, if your Farmout involves a single Drilling and Spacing Unit with no "split assignment" of horizons, or if the Farmee is to be assigned (and the Farmor will reserve) a uniform interest throughout the entire Farmout Area and a solid, "windowless" Lease Block is involved.*** You still may want to consider a Present Assignment/Sublease or Tax Partnership, though, if there are the additional complications of an intentional Off-Lease earning well, disproportionate sharing or Payout limitations, or problems concerning Delay Rentals or Percentage Depletion.

2. ***You should have FEW PRACTICAL PROBLEMS if relatively little acreage is involved, or the acreage value is expected to remain relatively low even if the earning well is successful.*** Again, consider a Present Assignment/Sublease or Tax Partnership only if the other tax considerations, listed above, are significant.

3. ***If aggregate Outside Acreage value might be high following the completion of a successful earning well, and a considerable amount of Outside Acreage is involved, you DO HAVE A PRACTICAL PROBLEM. Consider:***

(a) ***A Present Assignment/Sublease if the current***

aggregate Fair Market Value is low;[136]

(b) ***A Present Assignment/Sublease if there has been no increase in Fair Market Value, despite a high initial cost, and you are the FARMOR.***[137] **(If you are the Farmee, insist on a Tax Partnership.)**[138]

(c) ***A Tax Partnership if the acreage has already significantly increased in value.***[139]

All too often, no one "puts the pencil" to the problems under Revenue Ruling 77-176 and the related tax problems to determine whether or not serious tax consequences will result if a conventional Farmout structure is used. This particularly has been true of major oil companies, which have stayed away from Tax Partnerships in Recognition of Income situations if not insisted on by the other party. They forget that, if the major is the *Farmee,* or the other party is uninformed, hundreds of thousands (or millions) of dollars of unnecessary tax liability may result.

Equally unfortunate, most majors have insisted on Tax Partnerships in "Disproportionate Sharing" situations, without considering the dollars saved versus the dollars spent.

An *intelligent* Farmout game plan requires, instead, an *informed decision* concerning the practical consequences of each possible

---

136 But how likely is it that the earning well will be successful? If it is not, would the acreage be released? A Tax Partnership could still be a better solution (see § **3.04(b)[2][c-5-a]**. A Tax Partnership should also be considered if there are additional disproportionate sharing or Payout limitation problems. (The Tax Partnership should be coupled with a Present Assignment/Sublease if there are Delay Rental/Percentage Depletion problems.)

137 This assumes that you don't care what happens to the Farmee, and he is too dumb or ill-informed to insist on a Tax Partnership.

138 Again, coupled with a Present Assignment/Sublease, if there are Delay Rental/Percentage Depletion problems.

139 If there are also Delay Rental/Percentage Depletion problems, the Tax Partnership should be coupled with a Present Assignment/Sublease.

structuring alternative, including, where appropriate, a "Quick and Dirty" quantifying of potential results.

§ 3.04(b)[3] Structuring the Interests Assigned and Reserved

§ 3.04(b)[3][a] In General. The conventional Farmout structure has been discussed at length in preceding sections.[140] As we saw in **§ 3.04(b)[c-5-b-4]**, however, consideration of utilizing the Tax Partnership as a means of avoiding problems created by Revenue Ruling 77-176 paved the way for more flexible thinking on how the interests in production assigned to the Farmee, and retained by the Farmor, might be structured to more effectively accomplish the objectives of a particular Farmout.

The Farmout structure which you attempt to negotiate for your company will take into account your reasons for entering into the Farmout (see **§ 3.01(a)**) the economic impact of various Farmout structures (see **§ 1.03(c)[4]**), including the tax implications of the various alternatives available to you (see **§ 3.04(b)[2]** and its subsections); and your relative bargaining position *vis-à-vis* the other party to the deal (see **§ 1.03(d)**). Far too often, however, both the Farmor and the Farmee automatically accept whatever is the "standard" arrangement in the area, without considering structuring alternatives that might prove more beneficial to both parties. Likewise, relative bargaining position, not only as it affects the particular prospect but also as it applies to current industry conditions, is frequently ignored, particularly by the Farmor.

In the 1977-1981 period, parties who controlled valuable acreage positions found themselves in a true "sellers' market." Acreage was in short supply. In addition, drilling capital was readily available. Under these conditions, almost any type of deal, no matter how heavily "loaded," could be sold — at least to *someone*.

---

140 See § § **3.01(b)[4]**, **3.04(a)**, **3.04(b)[1]**.

In the 1982-1983 period, these conditions radically changed. Most companies, both majors and Independents, found themselves in a drilling capital "bind." Accordingly, the majority of majors became far less receptive to accepting outside deals, and instead used their more limited drilling budgets to attempt to save the most valuable portions of their own lease inventory. At the same time, those Independents who did have drilling dollars to spend found that they could pick and choose among very attractive prospects.

Under these circumstances, the Farmor who attempted to sell his deal on the same terms that had been readily acceptable in the 1977-1981 period found, to his surprise, that Farmees were no longer buying. Unfortunately, many Farmors continued to attempt to turn deals on the old "sellers' market" basis, wondering why valuable leases were going begging. Many valuable lease plays terminated, since no one could be found to drill the acreage. The more alert Farmor, however, simply accepted the fact that deal terms must be improved if acreage was to be saved. So, the "traditional" terms of a few years earlier — which, in fact, represented an anomaly brought on by the unprecedented demand for deals — were replaced by lower "promotes," the return of Outside Acreage as a Farmout sweetener, and, in some instances, deals being turned on a totally umpromoted basis.

The pendulum will again swing back, and will continue to gravitate between buyer and seller markets. The Farmor — or Farmee — who is alert to these changes will be able to structure deals which are appropriate to the times; other industry members will simply see deals not accepted, or will give (or take) deals on terms which are far less advantageous than could have been negotiated by more realistic analysis.

However, when unusual bargaining positions are not an overriding consideration, it is the *economics* — both tax and non-tax — that should control as to the deal structure selected in the absence of highly unusual conditions.[141]

The impact of these economic conditions on deal structure, and how they can be taken into account in structuring the deal, will be discussed in § **3.04(b)[3][c]**. However, just as it is impossible to determine whether or not theoretical tax considerations must be compensated for in deal structure, and what deal structure will be most effective in dealing with those consequences without running a simple "economic model" (see § **3.04(b)[2][d]**), an intelligent evaluation of deal alternatives requires a similar "Quick and Dirty" economic evaluation.

To determine how to best structure your Farmout to attain your company's goals, or to make your deal more attractive to a potential Farmee in "buyers' market" times, it is thus critical that you be aware of the various structuring alternatives available.

§ 3.04(b)[3][b] Available Structuring Alternatives. Interests assigned and reserved can include the following:

1. An *Overriding Royalty Interest* ("ORI") — a non-cost-bearing ("non-operating") interest measured out of *gross* production which will continue for the life of the lease.[142]
2. A *Net Profits Interest* ("NPI"), measured out of gross Working Interest revenue *less* costs of drilling, completing, equipping and operating the well (a "true" Net Profits Interest); or out of gross Working Interest revenues *less* operating costs (a *Net Operating Profits Interest* or "NOPI").

[141] An example of such unusual conditions would be the impact of *your particular objectives* in making a Farmout. For instance, if maintaining deep rights under an extensive lease block was the motivation for the Farmout, you might farmout the shallow rights on a "brother-in-law" basis in return for your reservation of deeper horizons and the Farmee's commitment to an aggressive shallow drilling program. See § **3.01(a)**.

[142] If a purported "Override" is measured out of something other than gross production, it is a *Net Profits Interest*, not an Overriding Royalty Interest. If it is a terminable interest, it is a *Production Payment*.

3. A *Production Payment* ("PP") — a *terminable* interest measured out of production which is limited to a specified *quantity* of production, a specified *dollar amount* of production revenues, or production received over a *specified period of time.*

4. A *Carried Working Interest* ("CWI"), which is non-cost-bearing *up to given point* (most commonly "to the casing point," but sometimes "through the tanks") but which immediately receives its share of Working Interest production.

5. A *Reversionary (Subordinated) Working Interest,* which shares neither in Working Interest costs nor Working Interest production until *Payout* — the point at which the Farmee has recovered, out of net Working Interest production, his costs of drilling, completing, and equipping the well.

6. A *Disproportionate Working Interest,* which bears a greater portion of Working Interest costs than the share of Working Interest production which it receives.

7. A proportionate *undivided Working Interest* ("WI"), where Working Interests costs, and Working Interest production, are borne and received in the same ratio.[143]

As we will see in § **3.04(b)[3][c]**, the typical Farmout has customarily involved a combination of the various structuring alternatives as to the interests assigned and reserved. Traditionally, as to the earning-well Drilling and Spacing Unit, one set of interests would be reserved by the Farmor, and assigned to the Farmee, *during the Payout period* ("BPO"), and a different set of interests *after Payout* ("APO").

A different "combination" arrangement also traditionally existed between the interests assigned and reserved in the *earning-well*

---

143 Traditionally, such "unpromoted" arrangements have been limited to Joint Operating Agreement (see **Chapter Four**) or Outside Acreage situations. However, when a true "buyer's market" exists, as during the 1982-1983 period, a Farmor may find it necessary to "turn" interests in his leases on a "ground floor" ("heads up") basis, simply to expand his drilling budget and to keep his leases from terminating.

Drilling and Spacing Unit and the interests assigned and reserved in the *Outside Acreage*. The Farmee would be "promoted" as to the Drilling and Spacing Unit on which the earning well was located (usually through the reservation of the combined convertible Override/Reversionary Working Interest), but would receive an unpromoted, *undivided* Working Interest in the Outside Acreage, with the Farmor also reserving an undivided Working Interest. In another alternative — to be discussed in § **3.04(b)[3][c]** — a "Checkerboard" arrangement would be utilized — the Farmee would be assigned 100% of the Working Interest in *alternate Drilling and Spacing Units*, with the Farmee reserving 100% of the Working Interest in the balance of the DSUs.

There is no magic that says that combination arrangements must be used in connection with Farmout structure. In fact, a common pattern in developmental Farmouts in a mature area is to assign the Farmee 100% of the Working Interest *throughout the entire Farmout area*, with the Farmor reserving a Non-Convertible Overriding Royalty Interest or Net Profits (or Net Operating Profits) Interest throughout the Farmout Area as a whole. However, as we'll again see in § **3.04(b)[3][c]**, varying the interests assigned and reserved as between the earning-well Drilling and Spacing Units and the Outside Acreage, and even *within* the earning-well Drilling and Spacing Units, is not only customary industry practice; in the majority of instances, it is also based on sound economic principles.

Unfortunately, considerable confusion concerning the nature of the interests assigned and reserved under a Farmout has been created by an indiscriminate use of terms, without insuring that the parties are using the terminology in the same manner.[144] While this is usually resolved, or at least discovered, at the time the Farmout Agreement is finalized, this failure to have reached a true "meeting of the minds" may at best cause a wild scramble to renegotiate the

---

144 The importance of clarity in contract terminology is discussed in **Chapter Eight**.

deal, or may cause a desirable deal to fall through. Even more unpleasant consequences arise when the agreement of the parties has been reduced to an unclear and incomplete "Letter Agreement,"[145] and the misunderstanding concerning the meaning of the terms used is not discovered until after drilling operations have commenced (or been completed!).

One area of potential confusion arises when the Farmor reserves a *Net Profits Interest* and/or *Net Operating Profits Interest.* The term *Net Profits* Interest is often used, indiscriminately, to describe both the "true" NPI, where the Farmor will not share in production revenues until Payout, and the NOPI, where only operating costs, and not the costs of drilling, completing, and equipping the well, are taken into account in determining whether or not any "net profit" exists.

Furthermore, while an *Overriding Royalty Interest* is a term of art, whose share of production will be calculated as a matter of law,[146] *Net Profits Interest* is not a word of art; the agreement must define at what point "Net Profits" (or "Net Operating Profits") actually occur.

Besides the obvious distinction between the true NPI and the NOPI, this requires a consideration, among other factors, of whether revenues are to be net or gross of production, severance, and excise taxes (including the Crude Oil Windfall Profit Tax); if an interest or

---

[145] As to the use of Letter Agreements, see § **1.01(c)**.

[146] While the Overriding Royalty Interest shares in "gross production" — free of *all costs* of drilling, completing, and equipping the well — and *most* costs of operation, it is not entirely cost-free. Those same costs that would be taken into account as a charge against a Lessor's share of royalty production — severance and production taxes, and, depending upon the wording of the reservation, costs of transportation, processing, and treating — will be used to reduce the proceeds distributable to the owner of the Override. Likewise, construction problems arise concerning whether the owner of the Override is to receive an "in kind" interest in production, or will be limited to receiving a monetary payment. Further problems develop in interpreting a "money payment" Override, due to the distinction between "proceeds" and "Market Value" royalties.

Time Value factor is to be taken into account as an "expense" item;[147] whether expenses are to be reduced, or revenues increased, by "Tax Benefit" considerations; whether, in the NOPI, a "depreciation factor" on equipment is to be taken into account, and, if so, how depreciation will be computed; and whether, in determining "costs," the Drilling Party will be entitled to take into account those same costs, and will measure those costs in the manner, set forth in a typical "COPAS" accounting procedure, including a consideration of whether or not an allowance will be made for those "indirect" costs provided for in Section III of the COPAS accounting procedure.[148]

Production Payments are less common in today's Farmout arrangements as a result of the Tax Reform Act of 1969. In certain circumstances, however, a party may still desire to use the Production Payment. If so, a problem also arises — where the Farmee is to be assigned less than 100% of the *entire* Working Interest in the acreage — as to whether or not the Farmor's stated interest is to be (or already has been) proportionately reduced. If an Overriding Royalty Interest is involved, this problem is limited to consideration of whether the expressed fraction was to be the gross or net interest, and also if it is to be received out of 100% of well production, out of the Working Interest share of production, or something in between. Where a Production Payment is involved, proportionate reduction questions involve not only whether or not the *share* of production out of which the Production Payment is to be measured is to be proportionately reduced, and is to be measured out of gross or net, but also if any *dollar* or *volume amount* is also to be proportionately reduced.

The *Carried Working Interest* often creates more problems concerning the actual intention of the parties than any of the other types of interest which a Farmor may reserve. If you and I agree that

---

[147] Time Value considerations are discussed in § § **1.03(b)[8]** and **1.03(c)[1]**.

[148] As to the COPAS accounting procedure, see **Chapter Four**.

you will "carry me for quarter," you may assume that my interest is to be subordinated to your recovery of your costs of drilling, completing, and equipping the well — in other words, that I am reserving a Reversionary Working Interest. This dual use of the term "Carried Working Interest" or "carry" creates the same unfortunate consequences that arise anytime one term is used indiscriminately to cover two totally different concepts.[149] To avoid this confusion, the use of the term "Carried Working Interest" should be limited to a true "carry" situation, in which the owner's participation in Working Interest production is *not subordinated* to Farmee Payout.

The distinction between the true Carried Working Interest and the Reversionary Working Interest is not the only cause of confusion concerning the Carried Working Interest. Another question is to *what point* the Farmor is to be "carried." In the typical "Third-for-a-Quarter" arrangement, the Farmee agrees to pay 100% of the costs incurred "to the casing point" — i.e., until the well has been drilled to total depth, testing completed, and a decision made as to whether or not to attempt to complete the well — in return for Assignment of a 75% Working Interest. In other words, the Farmor has been "carried to the casing point" for a 25% Working Interest. In another common variation of the Third-for-a-Quarter deal, the Farmor is carried "to the tanks," i.e., through the installation of production equipment.[150] Additional confusion arises concerning whether the 25% "promotion" also extends to previously-incurred lease costs; extends to all costs *actually incurred* in drilling to the point at which the "carry" will cease (or only to the AFE estimate of those costs); and whether the Farmor will also be permitted to reserve an Overriding Royalty Interest in an amount equal to the difference between existing

[149] A similar problem develops when there is an indiscriminate use of the terms "Working Interest Area" and "Area of Mutual Interest." See **Chapters Four** and **Seven**.

[150] Considerable confusion can also exist where an agreement has been reached to carry the Farmor "to the casing point" or "to the tanks," without any clear definition of these terms. See **Chapter Eight.**

burdens on the Lease and some specified *Net Revenue Interest,* e.g., 75%.

Even where the parties are in agreement as to the point of the carry and the other complicating factors mentioned above, a "meeting of the minds" has often failed when there has been no agreement reached as to whether the "carry" would apply *only to the first well drilled on the prospect* — the most common form of the Third-for-a-Quarter arrangement — or would extend to all wells.

It's important to distinguish the Carried Working Interest arrangement from the *Disproportionate Working Interest.* While a Carried Working Interest will, at some point, share in its proportion of all subsequently incurred costs, the "promotion" associated with the Disproportionate Working Interest is not so limited. Here, the promoted party bears a disproportionately lesser share of *all costs* associated with the wells (with the normal exception of monthly operating costs), rather than, at some point in time, bearing his proportionate part of subsequently incurred expenditures.[151]

We've now seen some of the structuring alternatives that are available to you. We've also seen some of the problems that exist when you and the other party to the deal are not using the terms in the same manner, or have not considered certain potential points of disagreement, so that there has not been a true "meeting of the minds." At the risk of unnecessary repetition, it cannot be emphasized too strongly that it is critical that you make sure, *while the deal is being negotiated,* that you are truly in "mutual agreement" with the other party concerning the deal terms. Otherwise, at the time of finalizing the "formal" contract, the deal may blow apart —

---

[151] In a *Disproportionate Working Interest* situation, the Farmor might agree to bear 25% of all well costs in return for a 50% Working Interest. Here, again, the parties normally share proportionately in monthly operating costs; however, costs of reworks would usually continue to be shared on a disproportionate basis.

A common variation of the "Disproportionate Sharing" involves a "staged" sharing, in which the percentage of costs borne by the promoting party will increase as costs exceed certain levels. See § **3.04(b)[3][c]**.

often when it is too late to make new arrangements to prevent lease termination; or the parties may find themselves in a lawsuit concerning the actual basis of the deal.

Now it's time to decide which of these various structuring alternatives will be most appropriate in structuring a particular deal!

§ 3.04(b)[3][c] Selecting an Appropriate Deal Structure. Most explorationists understand the nature of the various interests we have just discussed. However, many do not understand when it is to the advantage of their company to use one over another. Even more don't even seem to care — after all, aren't Farmout Agreements always structured in the same way?

Even where conventional Farmout structures are used, the *size* of interests assigned and reserved has tended to vary from region to region, as do the practice of committing an amount of Outside Acreage to the deal.[152] And even in conventional Farmout arrangements, the size of the interests assigned and reserved might vary if a particularly high-risk Farmout were involved, economics were particularly tough, or industry conditions or management's position required a variation from the more normal regional pattern.[153] Far too often, though — with the exception of such factors — little consideration has been given to whether an alternate deal structure was called for in a particular Farmout.

Intelligent Farmout negotiation requires that the structuring of the Farmout be "tailored" to the nature of the prospect and the objectives of your company and the other party to the deal, and that you further analyze the economics of the prospect to determine which alternative deal structure makes the most sense. Such a flexibility in thinking is particularly called for if unusual industry

---

152 As to *Outside Acreage*, see § **3.03(b)[6]** *et seq.*

153 As to the impact of various factors on Farmout structure, see § § **1.03(d)** and **1.04(a)**.

conditions create either a buyers' or a sellers' market. However, you can only utilize such creative structuring if you are aware of the advantages of the various alternatives available to you.

Let's look at a few examples. If a "Prospect Area" is marginal, or involves the likelihood of a single well field, reserving a *Convertible* Overriding Royalty Interest in the earning-well DSU, or undivided *Working Interest* in the Outside Acreage, makes little sense. Similarly, if your company is under extreme budgetary constraints, the reservation of an Overriding Royalty Interest or Net Profits Interest *throughout the entire Farmout Area* might make much more sense than reserving *any* type of Working Interest. In a like vein, if the *Farmee* has serious budgetary problems of his own, or would be an undesirable "partner" for future drilling, assigning undivided Working Interests in Outside Acreage is not the most sensible arrangement.

When you, as a Farmor, own Farmout Leases which are due to expire and have already been "Top-Leased," deal "sweeteners" may certainly be called for in your Farmout proposal, particularly if you have budget problems or find yourself in a buyers' market.[154]

The Overriding Royalty Interest, since it is non-cost-bearing, has the advantage to the Farmor that he need no share in, nor have his interest in production reduced by, any of the costs incurred in the conduct of prospect operations. (In other words, the Farmor receives *something* on *any* productive well, no matter how marginal.) Its main *disadvantages* are that: (i) the Farmee will normally insist that the Override be limited to a rather small fraction of production; and (ii) too large an Override may wreck the deal — cause the Farmee to elect not to complete an earning well, to prematurely abandon it, or not drill additional wells on the Farmout Leases.

A variation of the Overriding Royalty Interest which avoids the Override's normal disadvantages is the *Variable ("Flexible") Over-*

---

[154] Incentive Farmout structures are discussed in § **3.04(b)[3][d]**.

riding Royalty Interest, under which the *size* of the Farmor's interest in production varies with the success of the well.[155] In its most common form, the Farmor's Override increases in those months in which daily average production equals or exceeds a specified figure.

In drafting such a "flexible" provision, it is critical that the Farmout Agreement clearly specify whether the increased Override will apply "across the board," or if it is to be limited to the production above the "flash point" level. Other problems arise as to whether the "average production" will be arrived at well-by-well or on the Farmout Area average, and whether or not dry holes will be taken into account where average multi-well production is the measuring standard.

It is also critical, from the Farmee's standpoint, that the higher Override be payable only in months when the measuring wells have *actually produced* the higher figure, and not in those months when the wells were merely "capable of producing" the greater amount.

(And avoid the pitfall created by one Farmor's drafting error where his larger Override was based upon the specified monthly "barrels of oil" produced, and the earning well came in as a prolific gas well!)

In some instances, the Flexible Overriding Royalty, rather than involving an *increase* in the Royalty fraction when production exceeds a specified amount, involves a *reduction* in the fraction when unusual costs, such as the installation of a compressor or a pumping unit, must be incurred.

The Flexible Overriding Royalty Interest has advantages to both the Farmor and the Farmee. It permits the Farmor to receive a larger interest in production on those wells which can "handle" such a load, and protects the Farmee against a crippling "promote" on more marginal wells. It also avoids destroying well economics, with the resulting cessation of further drilling on the Farmout Area or

155 The Variable Override is a common arrangement in Canadian Farmouts.

abandonment of wells already drilled — and this benefits *both* the Farmor and the Farmee!

As mentioned above, customary practice in the United States is to reserve a relatively small Overriding Royalty Interest in the earning-well Drilling and Spacing Unit during the Payout period, with the Farmor reserving the option to convert his Override to a much larger Reversionary Working Interest After Payout ("APO").

There is no magic that says that the Override must remain relatively small during this Payout period; in Canada, the customary industry practice has been to reserve an Overriding Royalty Interest during Payout that is the estimated economic equivalent of the Farmor's After-Payout Working Interest. For economic reasons, however, the U.S. practice has normally been to retain the relatively small Before Payout ("BPO") Overriding Royalty Interest as a "sweetener" to the Farmee to permit him to recover his costs out of a rather large fraction of well production. As a *"quid pro quo"* for this arrangement, the Farmor is permitted to convert this Override to a relatively large Working Interest after the Farmee had recovered his costs of drilling, completing, and equipping the earning well.

As we'll see below, it would be equally possible for the Farmor to attain his larger interest in "APO" production by *increasing his Override* or by *converting to a Net Profits Interest.* Both of these arrangements would have the advantage of avoiding any APO cost liability. (And, if the increased Override were selected, any delay in the receipt of further revenues in the event costly reworks, blowouts, etc. put the well back into a "negative" position, would be avoided.) However, the major consideration (in the minds of most Farmors) that led to the choice of the Override convertible to a Reversionary Working Interest over other, non-cost-bearing structures was the desire to exercise *control over the well* at the earliest possible point in time.[156]

---

156 This same reasoning led to the virtually inevitable reservation of an *undivided Working Interest* in Outside Acreage. See below.

(Despite the desire for well control, the informed Farmor would make the conversion from Override to Working Interest status *optional* on his part, rather than automatic: there is no assurance that a 50% Working Interest will, in fact, equal or exceed in value the reserved Overriding Royalty Interest.)[157]

The "price" which a Farmor must usually pay for reserving a *Convertible* Overriding Royalty Interest is the reservation of a smaller interest in production than could normally be reserved if no right to convert were sought. Accordingly, if the Farmor anticipates that the Farmout Area wells may never payout, or will be somewhat marginal After Payout, he will usually be better off reserving (and normally *can* reserve) a larger Non-Convertible Overriding Royalty Interest.

If the Royalties reserved by Lessors have varied within a given Farmout Area, or Farmout Leases are subject to varying levels of *intermediate Overrides*, it becomes difficult to fix a "median" Override which the Farmor can reserve upon assigning the Farmout Leases without, on the one hand, overburdening certain leases, or, on the other hand, having other leases assigned to the Farmee on overly-favorable terms. To avoid this problem, a common practice is to reserve an Overriding Royalty Interest equal to the difference between some specified Net Revenue Interest and outstanding burdens.

At first blush, it might seem to be of little benefit for a Farmor to reserve a *Net Profits Interest* or a *Net Operating Profits Interest* rather than an Overriding Royalty Interest. The Net Profits Interest will not share in revenues until the Farmee has attained Payout. (Of course, as indicated above, and as discussed below, a Farmor could reserve an Override or an NOPI during this Payout period.) Even as to the Net Operating Profits Interest, since it is measured out of *net*

---

[157] A mandatory conversion might also result in the reserved "BPO" interest being treated, for tax purposes, as a Production Payment rather than as an Overriding Royalty Interest.

rather than gross production, it would at first glance seem to be a less favorable arrangement for the Farmor.

Bargaining considerations aside, the advantage of the NPI or NOPI is that its negotiation will usually permit the Farmor to reserve a *larger fraction of production* (albeit of the "net" variety) than would be the case if an Overriding Royalty Interest were being reserved. For instance, if a normal "trade" would be the reservation of a 1/16th Overriding Royalty Interest during the Payout period, a Farmor might be able to reserve a 25% Net Operating Profits Interest.

Which of these arrangements will be better from the standpoint of the Farmor (or the Farmee)? Obviously, it is impossible to know any absolute answer to this question until after the earning well has been drilled and the level of actual production is determined. However, by a discussion with your geologists and engineers, and a little work with your economics department (or some "Quick and Dirty" studies on your own), it is possible to reach an informed decision as to which of these interests is more likely to furnish you a greater share of production — assuming that the earning well performs as anticipated.

If it is expected that the earning well may be somewhat marginal, the reservation of the Overriding Royalty Interest — measured out of gross — will normally be in the Farmor's best interest. On the other hand, if the earning well is expected to be a relatively prolific producer, reservation of a larger Net Operating Profits Interest will better serve the Farmor.

(In the marginal situation, of course, you must also remember that the reservation of *too large* an Overriding Royalty Interest may cause the Farmee to elect not to complete the earning well, or to abandon it at an early date. So, an additional advantage of the Net Operating Profits Interest is its impact on the Farmee's decision to complete the earning well, or to continue to produce the earning well during periods of pipeline proration or temporarily low oil and gas prices.)

The "true" Net Profits Interest, payable only out of *After Payout* profits, has several advantages to the Farmor over the Reversionary Working Interest. The Farmor need not share in subsequent "rework" expenditures or other costs that may put a Working Interest owner into a subsequent negative cash flow position — temporary or otherwise. Thus, if the reworking of a well is unsuccessful, and the costs of the rework (or a deepening or plugging back) are never recouped, the Farmor who has reserved a Net Profits Interest, although not receiving any additional revenues, will not be forced to reach into his own pocket to pay his share of the loss. Since his right to receive revenues are merely deferred, and he is not put to any "out-of-pocket" expenditures, he also bears none of the Time Value risks. This is particularly advantageous to the Farmor, since, under current industry practices, no Time Value or "interest" factor is usually taken into account in determining whether or not *Net Profits* exist.

(Obviously, if you are the Farmee, you should insist on Time Value being taken into account in defining *Net Profits.*)

The disadvantage of the Net Profits Interest over the Reversionary Working Interest is that the Farmor has no control of the well. However, this supposed "advantage" of the Reversionary Working Interest may be overrated. Much the same benefits could be secured by imposing contractual obligations on the Farmee, or permitting the Farmor some voice in well decisions even though his interest is of the Net Profits variety. The Farmor can also reserve the right to operate the well, on a contractual basis, After Payout.

As to Outside Acreage, the same desire for control over the drilling of subsequent wells has led to the customary Farmout structuring, under which the Farmee is assigned an undivided Working Interest and the Farmor reserves an undivided Working Interest. However, as indicated above, if it can be expected that further drilling on the Farmout Area will not meet the Farmor's normal return standards, the reservation of an Overriding Royalty Interest throughout the entire Farmout Area may prove far more advantageous economically. Or, in an expected "profitable" area — and subject to previous comments concerning the importance (or lack thereof) of well control — the reservation of a Net Profits Interest would be the economic equivalent of retaining a similar undivided Working Interest.

(In fact, if Time Value is not taken into account in defining *Net Profits*, the NPI would be *more beneficial* to the Farmor.)

The reservation of the Net Profits Interest throughout the Farmout Area could also be greatly favorable to the Farmor if it is anticipated that all or a part of the additional drilling may be high risk in nature — e.g., where "one-well fields" are quite common in this type of prospect. Farmor risks will be even further reduced if Net Profits are determined on a well-by-well and not on a "whole prospect" basis.

Another arrangement which should be considered where Outside Acreage is involved is assigning and reserving acreage on a "Checkerboard" basis rather than assigning and reserving undivided Working Interests. Each of the parties will now be able to maintain personal control of their own acreage, including greater flexibility as to *when* their acreage should be developed. Likewise, either party can now "sit out" on a well being drilled by the other party without being subjected to a "Non-Consent" penalty.[158] This, at least in theory, gives a party another real advantage, in that he can "wait out" the drilling of an offset well by the owner of the adjacent acreage until the results of that well can be evaluated. If the Farmor would have joined in a well being drilled by the Farmee on one of the DSUs assigned to the Farmee, he can now drill "heads up" on his own offset. If he would not have joined in the well, he frequently can now enter into a new Farmout of his adjacent acreage to the party drilling the offset, and reserve a current interest in production, rather than be subjected to a Non-Consent penalty.

---

158 As to Consent/Non-Consent arrangements under Joint Operations, see **Chapter Four.**

Despite the benefits of the Checkerboard, such an arrangement also has its disadvantages. "Spreading" of your drilling budget has been eliminated, and both parties will be required to duplicate tank batteries and other "common" equipment and personnel that would have reduced the overall costs of drilling in the area. In areas where there is "no such thing as an offset" — in other words, where adjacent acreage may vary radically in well success — each of the parties is now under a speculative gamble as to which DSUs will prove more valuable. Likewise, if your acreage is the one with "short fuse" leases, it may be *you* that will be forced to drill prematurely and prove up the other party's acreage. (In any event, you may still be "stampeded" into drilling if a productive well on the other party's offset tract would drain your acreage.) Furthermore, a Checkerboard arrangement, being less common in the industry, is often viewed with suspicion by a Farmee, simply because he has a lack of familiarity with this kind of an arrangement.

So, there is no "right" or "wrong" answer as to whether or not a Checkerboard arrangement will prove in your best interest. Here, again, it is imperative that you analyze, from both a geological, engineering, and an economic standpoint the nature of the prospect, to see if this less common relationship as to Outside Acreage may prove favorable. However, when you know that joint operations with the other party to this deal have usually been fraught with disagreements and difficulties, a Checkerboard should be given serious consideration.

The factors illustrate how the economics, or the geological and engineering characteristics of a prospect, have an impact on structuring your Farmout arrangements. Many structuring alternatives, however, that could have proved beneficial to the parties have, in the past, been ignored by both the Farmor and the Farmee. This was based upon erroneous assumptions concerning the "imperative need" to follow the conventional structuring to avoid a loss of IDC deductibility on the part of the Farmee.

There is no question but that the economic impact of IDC

deductibility is a crucial factor in structuring most Farmouts. See **§ 1.03(b)[6]**.[159] However, prior to the issuance of Revenue Ruling 77-176, and its "eye-opening" effect on the consideration of alternate deal structures, most industry members did not realize that there were many deal structures available which would permit the parties to "have their cake and eat it too" — to assign and reserve interests which permitted the Farmee to deduct all of the Intangibles paid by him, but which might also better serve the economic objectives of the parties than the traditional Farmout structuring. While some of these arrangements do involve the formation of a Tax Partnership (with its admitted disadvantages — see **§ 3.04(b)[2][c-5-b-4]**), a number are available which do not require any *Partnership* formation.

As discussed above, there are definite economic advantages, aside from tax considerations, in the conventional Farmout structuring. The assignment of 100% of the Working Interest to the Farmee, subject to the reservation by the Farmor of a relatively small Convertible Overriding Royalty Interest during the Payout period, does permit the Farmee to receive the lion's share of well Revenues during the Payout Period, while his costs are still "at risk," in return for the Farmor receiving a larger share of well revenues once Payout

---

[159] If this is the case, why was the use of the "Third-for-a-Quarter" arrangement — which clearly required capitalization of 25% of a drilling party's Intangibles under "conventional" structuring — so prevalent among Independents? First, non-tax economic considerations may always outweigh tax factors. From the standpoint of the Independent, the Third-for-a-Quarter arrangement permitted him to recoup his costs of putting a prospect together, and to participate in the well after the highest-risk expenditures — the costs of drilling to the casing point — had been incurred. Tax problems were frequently "avoided" by structuring the deal on an allegedly "heads up" basis, but utilizing an artificially inflated "*Turnkey*." (As to "*Turnkey*" drilling, see § **1.03(b)[6]**.) Furthermore, until recently, many Farmors (and Farmees!) did not realize that the *Carried Quarter* required the Farmee to capitalize 25% of his Intangibles. (And until the issuance by the Internal Revenue Service of its "Oil and Gas Audit Guidelines" in 1981, many IRS auditing agents were also unaware of this capitalization requirement.)

has been attained.[160] For this reason, even informed Farmors and Farmees may wish to structure the majority of their deals on the "conventional" basis.

Let's look, though, at some of the arrangements that are available to the parties as alternatives to the conventional structuring.

Even *without the formation of a Tax Partnership*, here are some of the ways in which your deal could be structured:

1. The Farmor can consider reserving an *NOPI* or a *Variable ("Flexible") Overriding Royalty Interest* during the Payout Period, rather than the "conventional" Convertible Overriding Royalty Interest. (He can also consider reserving a larger *Non-Convertible Overriding Royalty Interest*, as discussed above, rather than a smaller Convertible Override.)
2. Where an Overriding Royalty Interest is reserved during the Payout Period, the Farmor can reserve the right to *increase his Overriding Royalty Interest*, or *to convert to a Net Profits Interest*, after Payout rather than converting to a Working Interest. While the Farmor, at least in theory, will lose certain After-Payout "control" as to the earning well, he will, as discussed above, also avoid the cost burdens associated with Working Interest ownership. (Also, "control" over the well may be less significant in an earning well After-Payout situation than it would be as to the decisions concerning additional wells drilled on the Outside Acreage.)[161]

---

[160] In theory, the Farmee would not have been "made whole," despite this subordination of the Farmor's larger production interest, since Payout does not normally include Time Value considerations. However, the Farmee would often more than "break even," since *tax benefits* are also normally not taken into account in determining when Payout is achieved.

As to the defining of *Payout*, see **Chapter Eight**.

[161] Of course, there will still be certain major decisions concerning such questions as reworking, deepening, or plugging back the earning well. See **Chapter Four**.

3. The Farmor may wish to reserve *multiple options* at the Payout point: for instance, a right to convert to a Working Interest, *or* to a Net Profits Interest, *or* to a larger Overriding Royalty Interest. This would give the Farmor much greater flexibility in deciding what arrangement would be in his best interests after the well has been on stream for a sufficient period of time for him to make, hopefully, a more informed decision.

The rights just mentioned could be reserved by the Farmor without the formation of a Tax Partnership, and without causing any loss of IDC deductibility to the Farmee. If the formation of a Tax Partnership were necessary for other reasons, or the advantages to be gained under structuring alternates outweighed the disadvantages of filing the annual Information Tax Return required of Tax Partnerships (see **§ § 3.04(b)[2][c-5-b-4]** and **3.04(b)[3][d]**), there are *additional arrangements that can be structured without loss of IDC deductibility by using the Tax Partnership:*

1. The Farmor can *"promote" his well on a "Third-for-a-Quarter" basis, or otherwise retain a Carried Working Interest,* so that he can share immediately in well production. Such an arrangement can be extremely beneficial where, once testing has been completed, the completion decision is relatively risk-free.

   The disadvantages of the Carried Working Interest are that the Farmor *will* share in those completion risks — and don't let anyone tell you that such risks are always (or even usually) minimal or non-existent — and that the size of the Carried Working Interest which can normally be negotiated is usually smaller than the "APO" Reversionary Working Interest a Farmor could retain in addition to his small pre-Payout Override.
2. A *Casing Point Election ("Option")* can be reserved which will give the Farmor the option of electing, after testing has been completed, to share in future well costs and receive an

immediate share of Working Interest production—or, if he would prefer, to receive the smaller Convertible Overriding Royalty Interest.

The Casing Point Election gives the Farmor much greater flexibility than the Carried Working Interest, and usually does not require that he receive any smaller Reversionary Working Interest (if that option is elected) than could have been retained if the Casing Point Election had not been negotiated. However, the "carry" will still usually be smaller than the Reversionary Working Interest that could be received by subordinating to Farmee Payout.

3. A *Disproportionate Working Interest* can be reserved, under which the Farmor's promotion will usually apply to *all wells* and *all costs* (at least other than normal monthly "lifting costs.") If there is a high potential for cost overruns on the well, the Farmor may agree to a "staged" arrangement, in which he bears no portion of well costs up to a given cost level; a disproportionate part of subsequent well costs up to another designated level; and, from that point on, shares in further well costs on a proportionate basis. This "staged" Disproportionate Sharing gives the Farmor the advantage of a greater interest in production during the Payout period, in return for his agreeing to bear a portion of the "cost overrun" exposure. (This frequently permits a Farmor to "sell" a deal which otherwise might be unacceptable to the Farmee.) Normally, a Disproportionate Sharing arrangement permits the Farmor to receive the same Working Interest, again from "day one," as he would have received in a "Subordinated" position in return for his having provided the cost overrun "insurance."
4. A variation of Disproportionate Sharing is the *Tangible/Intangible ("Functional Allocation") arrangement*, under which the Farmor, in return for bearing all "Capital (non-deductible) Costs," receives a share of well production

which is significantly greater — normally 15 to 25 percentage points — than the "spread" between "Capital" and "Non-Capital Costs."

Unlike the Disproportionate Sharing arrangement, the Farmor in the Tangible/Intangible arrangement does not bear *any portion* of non-deductible expenditures; whereas in Disproportionate Sharing, the Farmor bears his specified portion of *all costs* "across the board."

Under such an arrangement, there is usually a "Special Allocation" of tax deductions so that the Farmee, whose expenditures have been limited to deductible costs, attains 100% deductibility for Federal income tax purposes. Such a structure may be particularly attractive when the Farmee is an Independent who is securing his drilling dollars from outside investors. (See § § **3.04(b)[2][c-1]** and **3.04(b)[2][d]**.)

The Tax Partnership will also permit the Farmor, if he so chooses, to put some limitation on Farmee *Payout*. The typical "Subordinated" arrangement may result in a substantial — and potentially unjustified — deferral of any significant interest to the Farmor in earning-well production in those circumstances where the Farmee has "loaded" his costs, incurs unjustifiable costs overruns, or sells his production to an affiliate at "brother-in-law" prices. (See § **3.04(b)[2][c-4-b]**.) Any Payout limitation would normally cause the Farmee to lose a substantial part of IDC Deductibility.[162] However, such Payout limitations can be utilized, without any adverse tax consequences to the Farmee, so long as a Tax Partnership structure is selected. Likewise, the Tax Partnership format will permit structuring the Farmout on an *Aggregate Well Payout* basis — prohibited, for tax reasons, under Revenue Ruling 80-109 — without any negative tax consequences. See § **3.04(b)[3][d]**.

---

162 The loss might occur even if the Payout limitation never took effect. See Revenue Ruling 80-109, and the discussion at § **1.03(b)[6]**.

The majority of the structuring arrangements discussed above would have required the Farmee to capitalize a portion of his Intangible Drilling and Development Costs if it were not for the Tax Partnership. However, "Special Allocations" of tax deductions in different ratios than the Pre-Payout Working Interest can be made where the Farmout is structured as a Tax Partnership, so long as the deductions are merely being allocated to the person who bore the expenditures which gave rise to the deductions.[163]

While the additional costs associated with the annual Information Tax Return required of a Tax Partnership must be balanced against the advantages of these alternate arrangements (see **§ 3.04(b)[2][c-5-b-4]**), use of the Tax Partnership often permits more advantageous economic arrangements — certainly from the standpoint of the Farmor — at no "tax" cost to the Farmee.[164]

Another structuring variation which can prove of great benefit to the Farmor when he is in a strong bargaining position is the *Option to Participate.* Under this arrangement, the Farmor, rather than merely reserving an undivided Working Interest in the Outside Acreage, reserves the *right to elect* either to participate in subsequent drilling under the Joint Operating Agreement or to receive a Convertible Overriding Royalty Interest on subsequent wells. The Option to Participate thus permits a Farmor who feels it is premature to participate in additional drilling, or may be convinced that the additional wells lack profit potential, to receive an immediate share of production under his Convertible Override rather than accept a penalty-burdened "Non-Consent" position under the Joint Operating Agreement as to such subsequent wells. (See **Chapter Four.**)

The Option to Participate is normally exercised at the time the

---

[163] See § **3.04(b)[2][c-5-b-4]**.

[164] To avoid the partial loss of tax benefits, however, the Farmout Agreement must be carefully drafted. See § **3.04(b)[2][c-5-b-4]**.

subsequent well is commenced on the Outside Acreage. The Farmor will attempt to reserve such an Option on a well-by-well basis on each subsequent well. However, most Farmees, if "willing" to agree to such a reserved Option (usually because of the strong bargaining position of the Farmor), will insist that the Option be limited to the first subsequent well, and that such election be binding as to all subsequent drilling on the Outside Acreage.

The Option to Participate is not a "plain vanilla" type arrangement. As indicated above, the Farmor must be in a strong bargaining position to be able to negotiate it. But it does give valuable flexibility to the Farmor when he finds himself in such a bargaining position.

Will these "unconventional" techniques work to your company's advantage? To answer that question, you must first be aware of the nature of the prospect and your geologist's or engineer's estimate of the probable drilling results. If a "bonanza" can be expected if the earning well is productive, then to accept the smaller After-Payout interest in production that a Carried Working Interest, Non-Convertible Override, or Casing Point Election would normally generate, will prove a disadvantage to a Farmor. Similarly, a Disproportionate Sharing arrangement may not be to your advantage where the dry hole risk is quite high, unless this is the "price tag" which must be paid to negotiate the deal. (On the other hand, this is exactly the kind of arrangement you would try to negotiate under such circumstances if you were the Farmee.) Thus, to understand when, and under what circumstances, to vary from the conventional structuring requires an understanding of the prospect, plus some type of economic evaluation — whether a sophisticated study by your economics department or a "Quick and Dirty" study you have scratched out for yourself — to see which of the arrangements is more likely to be in your favor.

Rules of Thumb:

1. ***If you are the Farmor:***

   a. ***A Non-Convertible Overriding Royalty Interest or***

Net Operating Profits Interest throughout the entire Farmout Area will be in your best interests if the prospect appears marginal: you'll normally be able to reserve a large fraction of production, not reduced by operating costs. (But don't make the interest so large that you kill off the deal.)

The decision between the ORI and NOPI should be based on just *how marginal* the prospect should prove, and the various production fractions that could be reserved. In other words, we can't estimate whether 1/8th of *gross* will be better than 25% of *net*, or vice versa, without running some estimated economics.

b. ***If the outcome of the drilling is speculative, consider a Variable Overriding Royalty Interest, or some Net Profits arrangement,*** particularly if reworks or the like could occur After Payout. Also ***consider multiple conversion options at Payout, or an Option to Participate.***

c. ***If PAYOUT is LIKELY, but the wells will probably NOT be SPECTACULAR, go for a "carried" arrangement*** — "Third-for-a-Quarter," Casing Point Election, etc. The larger interest in *immediate production* during the Payout period should prove more valuable than the greater reversionary interest. Also, run the economics of ***Disproportionate Sharing***, particularly if the dry hole risk is slight; or consider a ***ORI that increases After Payout.***

d. ***If the well, IF PRODUCTIVE, should be a "BARN BURNER," go for a larger Reversionary Working Interest,*** particularly where the risk of a *productive-but-disappointing* well is *slight.*

e. ***If HIGH RISK COMPLETIONS must be made to evaluate probable well success, try to stay away from "carries,"*** unless you can retain *a very high "carried"* *position* (and can afford the completion risks!), or can be *carried "to the tanks." **If wells must be PRODUCED for a substantial period of time to EVALUATE RESULTS or if many DRY HOLES, along with producers, are expected on the prospect, consider a Reversionary Working Interest coupled with an NOPI; or, better still, a well-by-well Net Profits arrangement.***

2. ***If you are the Farmee:***
   a. Obviously, ***the reverse of each of the foregoing is true.*** Also, carefully run *risk-adjusted economics* to ascertain whether that "sure thing" productive well really should be drilled at all, at least on the terms proposed.
   b. ***Where dry hole and cost overrun RISKS are HIGH, consider Disproportionate Sharing.*** You wouldn't be drilling such a well if you didn't expect a high multiple Payout if the well is productive. You are giving up a larger interest in Payout period production only, and are receiving a valuable "insurance policy."
   c. ***In MEASURING "PAYOUT," and particularly in Net Profits, seek an allowance for Time Value considerations.***

§ 3.04(b)[3][d] Incentive ("Sweetener") Arrangements. In **§ 2.03(b)[4]**, we discussed "incentive" arrangements in Support Agreements. The use of acreage and Depths Earned "sweeteners" was also touched upon in § **3.03(b)[6][c]**. (And, see also **§ 3.03(b)[4][a]**.)

Similar incentive arrangements can be used in connection with interests assigned and reserved to encourage a Farmee to accept a high-risk Farmout, to perform maximum "optional" drilling, or to drill to deeper depths. For instance, a Farmout Agreement can reduce the size of a Farmor's reserved interest during the Payout Period; the size of the Farmor's subordinated *After Payout* Interest;

or increase the size of the interest to be assigned to the Farmee in Outside Acreage (as well as the amount of that Outside Acreage) — *all IF the Farmee elects to test certain very deep formations, or to drill a larger number of optional earning wells.*

One of the most significant uses of "incentive" arrangements in connection with the Farmout arises during periods when deals are difficult to turn on any terms. When such conditions exist, reducing a Farmor's normal promotion to take into account the "buyers' market" conditions may not, by itself, be enough — it may also be necessary to insert some highly unusual provisions in the Farmout proposal to catch the potential Farmee's attention and to focus his eyes on the fact that this is a truly unusual (and favorable!) proposal.

The types of "sweeteners" which can be used are limited only by the Farmor's creativity. Alternatives which have successfully been used to catch a potential Farmee's attention include the following, among others:

1. Reservation of a *small* Non-Convertible Override only, with no After-Payout reversion as to the earning-well DSU. For this to be a meaningful incentive, the reserved Override would need to be small enough so as not to be simply the economic equivalent,of the interest that the Farmor would have received as an APO Reversionary Working Interest or the typically larger Non-Convertible Override. (See **§ 3.04(b)[3][c]**.)
2. Waiver of *any* interest in earning-well production during the Payout Period.
3. Reservation of an *unusually small* Reversionary Working Interest in the earning-well Drilling and Spacing Unit. Here, unlike the usual arrangement, the Farmor's Reversionary Working Interest would be smaller than the undivided Working Interest reserved in Outside Acreage.
4. Subordinating the right to convert to recoupment of 200% (or some other multiple) of the Farmee's costs, or allowing a Time Value factor.

5. Imposition of a "penalty" if the Farmor in a "Casing Point Election" arrangement elects not to participate in the earning-well completion attempt.
6. Conditioning the Farmor's right to convert in multiple earning-well Farmouts to *aggregate* earning-well Payout.[165]

The foregoing merely illustrate the varied arrangements which a Farmor may propose to focus the Farmee's attention on *his* deal as being an unusually fair proposal. Any arrangement which accomplishes the same result will do the trick.

§ 3.04(b)[4] Acreage and Depths Assigned. The interests to be assigned to the Farmee, and reserved by the Farmor, obviously must also take into consideration how much acreage is to be committed to the Farmout; how much acreage will be earned for each earning well drilled; and to what depths such acreage will be assigned. These topics have been discussed elsewhere in this chapter: (§ **3.02(b)**, **3.03(b)[5]** and its subsections; and § **3.03(b)[6]** and it subsections.)

§ 3.04(c) Operational Reservations. Obviously, since it is the Farmee's dollars that are at stake in the drilling of the earning wells, the Farmee will want to have control over the day-to-day operation of these wells which his dollars are paying for.[166] However, since the Farmor has committed his acreage to the deal, he will want some say concerning the conduct of the operations, and particularly over the future of the Farmout Leases in the event the Farmee elects not to complete the earning well or desires to release his interest in the acreage. Accordingly, the Farmout Agreement will usually contain

[165] Such an arrangement must be structured as a Tax Partnership, with "Special Allocations," to avoid negative tax consequences to the Farmee. See Revenue Ruling 80-109 and § § **1.03(b)[6]**, **3.04(b)[2][c-5-b-4]**, and **3.04(b)[3][c]**.

[166] Reservation of excessive control over the well might also make the Farmor liable to creditors or for third-party damages under a "Mining Partnership" theory.

certain "Operational Reservations" in addition to the interests reserved in production and acreage.

Certain Operational Reservations tend to be included in a Farmout Agreement as a matter of course and are not usually a subject of any extensive negotiation. These include the assignment by the Farmor of a right to take over operations if the Farmee elects not to complete the earning well (a *"Completion" Take-Over Right*); the right to take over the earning well to conduct his own tests (a *"Testing" Take-Over Right*); and, customarily, the right on the part of the Farmor (and, in some instances, a similar right on the part of the Farmee) to require notice and a *right to reassignment* of the Farmout Leases if the Farmee subsequently elects to abandon his interest in those leases.[167]

Certain other Operational Reservations on the part of a Farmor are not automatically included in a Farmout Agreement and are a subject for negotiation between the parties. These include a reservation of *processing rights* to extract liquids from any "wet" gas produced from the Farmout Leases; a *Call on Production*, i.e., the right of the Farmor to purchase the Farmee's interest in production either on pre-specified terms or at current rates in the field; and a *preferential right to purchase* the Farmee's interest in the Farmout Leases if the Farmee wishes to sell that interest.[168]

In addition to these "standard but negotiable" Operational Reservations, a given Farmout may call for less-standard provisions concerning Operational Rights, such as a right to consultation and *approval of any spacing application*, or the *right to take over operation* of wells on the Farmout Area once the earning wells have been drilled.

---

[167] A less-common reservation is a right of reassignment in the event of expiring leases. See **§ 3.03(b)[7]**.

[168] In some instances, the Farmee may also attempt to negotiate a call on the *Farmor's* share of production; his *own* right to process gas produced from the Farmout Area; or a preferential right to purchase as to the *Farmor's* Leases.

Just as problems exist in all areas of Farmout negotiation and drafting, certain pitfalls must be avoided in negotiating and drafting the Operational Reservations. Where, as is usual, a "Completion" Take-Over Right is reserved by the Farmor, it must be determined whether this Take-Over will extend as to all depths, or, as is more common, will be limited to the formation in which the completion attempt is made. The agreement should also specify whether or not the interest which the Farmee otherwise would have earned in that formation will revert to the Farmee if the completion attempt is not successful or the Farmee subsequently abandons the well in that zone. The "Completion" Take-Over Right may also create particular problems where there are multiple zones in which the earning well could be completed.[169]

Just as the Farmout Agreement should indemnify the Farmor against liabilities arising from the Farmee's operations (see § § **2.06(b)** and **3.05(d)**), the Farmee should be indemnified against liabilities arising during a Farmor Take-Over period.

Calls on Production primarily arise when one party to a Farmout is a major oil company (whether Farmor or Farmee) and the other is an Independent. (Need we say which party will be insisting upon the reservation?)

Aside from questions concerning the price at which such production can be purchased, additional problems develop when such a reservation could result in the arrangement being taxed as an "Association Taxable as a Corporation."[170] The Internal Revenue

---

[169] Similar Take-Over problems in connection with Support Agreements are discussed in § **2.05**. Problems concerning priority among completion and "split" ownership of horizons are discussed in **Chapter Four**. (See also **Chapter Four** concerning changes of ownership where less than all parties elect to complete a well, or where such a well is subsequently abandoned.)

As to all other less-common Take-Over provisions, — the right of the Farmor to drill deeper when drilling to *Contract Depth* is excused — see § **3.03(b)[4][c]**.

[170] Associations Taxable as Corporations are discussed in § § **1.03(b)[2][b]** and **3.04(b)[2][c-5-b-4]**.

The right to take production in kind under Joint Operating Agreements is discussed in **Chapter Four**.

Service has not, as yet, taken any firm position on whether the reservation of such a "Call" would have the dreaded "Association" result; and most industry members have elected to "sweep the problem under the rug" until it is raised by the IRS. Some Independents, however, refuse to grant a Call on Production because of these tax fears.

Many major oil companies insist on a Call on Production when dealing with an Independent as a matter of course. Periodically, a deal will fail because of a major's insistence on the Call, without any consideration being given to whether or not the Call on Production is of sufficient value to "blow" the deal, particularly in *this area.* Accordingly, before a Farmout is permitted to founder on this point, the value of receiving — or giving — the Call should be realistically evaluated.[171]

§ 3.05 Other Farmout Agreement Provisions

§ 3.05(a) In General. While the Subject Matter of the Farmout, the Earning Conditions, and the Interests Assigned and Reserved are the areas most commonly focused upon in Farmout negotiations, there are a number of other subjects that must be covered in the Farmout Agreement to round out its terms.

Some of these provisions — such as Insurance, Assignability, Legal Effect and Notice provisions (see **§ 3.05(d)**) and provisions concerning Title (see **§ 3.05(b)**) — are looked at as being so "boiler plate" in nature that they are neither discussed in the Farmout negotiations nor cause much stir when the formal Farmout Agreement is submitted. Other provisions, however, such as Delay Rental reimbursement (see **§ 3.05(b)**), and terms of the Assignment and any implementing Joint Operating Agreement (see **§ 3.05(c)**), are often ignored in negotiations, but may cause serious, last minute

[171] Granting a Call on Production can also be a Farmout "sweetener." See **§ 3.04(b)[3][d]**.

disagreement when the formal Farmout Agreement is submitted. Even worse, if the Farmee fails to carefully read the formal Farmout Agreement when it is submitted, or if the agreement is not formalized until after the earning well is commenced (or even completed), very real problems can arise.

The problems involved in commencing operations prior to formalizing the Farmout Agreement, or in submitting the formal Farmout at the last minute (when it may be too late to resolve potential disagreements) has been discussed in § **3.01(c)**. The problems that will be faced by a party who signs an agreement without having carefully reviewed its contents were highlighted in § **1.01(e)**. The agreement areas discussed in the following subsections may seem — and probably are — relatively insignificant when compared with the negotiation topics discussed earlier. But this will be of little consolation to the parties where the pitfalls highlighted in the balance of this Chapter create disputes that derail a potential Farmout or bind one party to an "agreement" to which he, in his mind, had never agreed.

§ 3.05(b) Title and Delay Rental Matters. It is unusual for a Farmout Agreement to contain Warranties of Title on the part of the Farmor. A "general" Warranty of Title — where the Farmor "guarantees" the validity of the Title to be acquired by the Farmee — while not absolutely unheard of, is highly unusual in a Farmout. A Farmout Agreement will occasionally contain a "special" Warranty of Title — the Farmor will guarantee that *he* has done nothing to create additional burdens on the Farmout Leases. However, the Farmout Agreement commonly contains *no* Warranties of Title, either general or special, and may even expressly negate such title guarantees.

Since the Farmor will not be warranting the title acquired by the Farmee, it is imperative, from the Farmee's standpoint, that sufficient time be allowed prior to the specified well commencement date for the Farmee to complete his examination of

Title.[172] The Farmor will normally not be required to deliver complete Abstracts of Title to the Farmee, but will be required to turn over any Abstracts and other title papers in his possession.

If the earning well is a *mandatory* well,[173] the Farmout Agreement should provide that acceptable title is a condition of the Farmee's obligation to drill.

Once the earning conditions have been satisifed and an Assignment is tendered to the Farmee (see **§ 3.05(c)**), the Farmee should run a "record check" to insure that no interim judgments have been rendered, or other liens created, against the Farmor; and that the Farmor has not created any intervening encumbrances or otherwise clouded the title that is being assigned to the Farmee.

In periods where the industry is in economic "trouble," it is very possible that actions may be taken against the Farmor during the earning period which could create liens against his title. While it is not customary to record Farmout Agreements, and the Farmee's activities on the Farmout Leases may well create a "duty to inquire" which could make the Farmee's title prevail, the Farmee's title would at best be clouded under these circumstances. Accordingly, if the Farmee has any concerns and wishes to reflect his rights of record, filing an *Abbreviated Summary Contract*, which merely recites that the Farmor and Farmee have entered into a Farmout Agreement, and incorporates the terms of that Agreement by reference, may be a wise protective step.[174]

---

172 Well commencement provisions are discussed in **§ 3.03(b)[1]**.

173 As to *mandatory* versus *optional* earning wells, see **§ 3.03(b)[3]**.

174 Filing the Summary Contract will protect the Farmee's rights without proclaiming to the entire world the terms on which the Farmout Agreement was entered into. (Of course, it will also require that any subsequent title examiner request a copy of the Agreement!) It is also a safer practice, from a legal standpoint, than merely filing an *Affidavit* concerning the Farmee's rights: under the laws of many states, Affidavits are not entitled to be received for record and will not serve as constructive *record* notice, even where they have been spread on the record.

For protection of both the Farmor and the Farmee, the Farmout Agreement must specify who will take responsibility for the payment of Delay Rentals affecting the Farmout Leases, and should further provide how such payments are to be borne between the parties. Since the Farmor will already have set up his records, the Farmout Agreement will customarily provide that the Farmor will continue to administer the leases and physically see to it that Delay Rentals necessary to maintain the Farmout Leases are paid. Almost invariably, though, the Farmout Agreement will excuse the Farmor from any monetary liability for a "missed" rental where such failure to pay is inadvertent.[175] The Agreement should further require that the Farmor submit proof of payment to the Farmee sufficiently in advance of rental dates so that any mistake in payment can be corrected.

The Farmor will normally physically administer the leases. But this does not mean that the *economic burden* will necessarily, or even customarily, be borne by the Farmor as well. Almost without exception, the Farmout Agreement will require the Farmee to reimburse the Farmor for all or a part of the Delay Rental payments made during the earning period.

The percentage of Delay Rentals to be reimbursed is a subject of negotiation between the Farmor and the Farmee. The most common practice in the United States is for the Farmee to reimburse the Farmor for 100% of such Delay Rentals, even though the Farmee will be earning less than 100% of the Working Interest in the Farmout Leases.[176] Once the earning conditions have been satisfied, future Delay Rentals will normally be borne by the parties in the same ratio

---

[175] *Cf.* liability in connection with payment of Delay Rentals and Shut-in Royalties under Joint Operating Agreements, discussed in **Chapter Four**.

[176] In some areas, the reimbursement is based on a pro-ratio of one year's rentals on the Farmout Leases. More commonly, though, the reimbursement is on an "as paid" basis, with the Farmee reimbursing the Farmor for the agreed percentage of Delay Rentals *actually paid* during the earning period.

in which they now own the Farmout Lease Working Interest.[177] In some instances, however, a Farmee may successfully insist that the Farmor, even during the earning period, be reimbursed only for the Farmee's proportionate part (rather than 100% of the earning-period Delay Rentals.)

Where customary Delay Rentals of one dollar an acre are "at stake," the question of the percentage of Delay Rental reimbursement by the Farmee may be more of a "one-upsmanship" game between the individual negotiators rather than a matter of true economic significance to the parties. However, where higher Delay Rentals are payable, such as is the general rule in Southern Louisiana, pro-rated versus 100% reimbursement of earning-period Delay Rentals may have a significant economic impact.[178] Similarly, when attention-getting "sweeteners" are being added to a Farmout proposal during "buyers' market" periods, the Farmor's negotiator may be wise to voluntarily limit the Delay Rental reimbursement requirement to the Farmee's pro-rata interest in the Farmout Leases; or, assuming the customary dollar-an-acre Delay Rental, to "graciously" agree that the Farmor himself will bear all Delay Rentals during the earning period.[178]

Occasionally, and particularly where multiple mandatory earning wells are required (see § **3.03(b)[6][c]**), Shut-in Royalties may also be payable during the earning period. Here, the same problems will arise concerning the administrative payment of the Shut-in Royalties; how the economic burden of such payments is to be

---

[177] Here, again, such rentals will normally continue to be paid by the Farmor, with the Farmee being billed for his proportionate part.

[178] As discussed in § **3.04(b)[2][c-4-b]**, when Delay Rentals are reimbursed prior to the formal assignment of the Farmout Leases to the Farmee, the Farmee will be required to capitalize these payments. Use of a *Present Assignment*, rather than the conventional *Assigned When Earned* approach, avoids this result and permits the Farmee to deduct the Delay Rental reimbursements. See § **3.04(b)[2][c-5-a]**.

[179] As to incentive arrangements in the Farmout, see § **3.04(b)[3][d]**.

allocated among the parties; and exoneration of the paying party for any inadvertent mispayment. Unlike the normal practice where Delay Rentals are concerned, it is more customary for the Drilling Party to administratively pay the Shut-in Royalties, billing the Farmor for his agreed-upon portion of the payment.[180]

§ 3.05(c) Implementing Provisions and Exhibits. The Farmout Agreement should contain provisions for implementing the Farmor's obligation to assign the Farmee his interest in the Farmout Leases once the earning conditions have been satisfied, and for the conduct of subsequent Joint Operations by the parties. Section **3.01(b)[5]** and its subsections outlines the pros and cons of attaching copies of a formal Assignment and a verbatim Joint Operating Agreement as exhibits to the Farmout Agreement, rather than merely inserting a "general" assignment provision and a reference to the Joint Operating Agreement form to be used. See also § **3.04(b)[2][c-5-a]** as to the advantages and disadvantages of utilizing a Present Assignment/Sublease, under which the Farmee will be assigned his interest in the Farmout Leases as of the date on which the Farmout Agreement is entered into, rather than the Assigned When Earned approach used in the "conventional" Farmout.

Some Farmout Agreements obligate the Farmee to formally request an Assignment within a specified period of time following the completion of the earning well. Under the terms of the Agreement, if such a formal request is not made, the Farmor has no obligation to assign the Farmout Leases to the Farmee; and the Farmee earns no interest in those leases, even though the other earning conditions have been fully satisfied. While there may be some question concerning the enforceability of this provision, the prudent Farmee will make sure that he has formally requested his Assignment, and that the Assignment is in fact received. Likewise,

---

180 Following the earning period, the payment of Shut-in Royalties and Delay Rentals will normally be covered by the Joint Operating Agreement. See **Chapter Four**.

when such a provision is contained in the Farmout Agreement, a "tickler file" reminder is certainly called for.

Often a Farmee accepts the Assignment as tendered without bothering to review the instrument for accuracy. If the Assignment contains provisions which are contrary to the provisions of the Farmout Agreement, omits Farmor obligations which should survive the earning period, or inserts new provisions not previously discussed, such as a Call on Production, the Farmout Agreement might be "merged" into the Assignment, and the Farmee bound by his acceptance of the Assignment. Accordingly, the instrument should be carefully reviewed for its legal sufficiency and its conformity to the Farmout Agreement before it is formally "accepted" by the Farmee or filed for record.

As to the common provision in the Farmout Agreement concerning the assignability of the Farmee's interest during the earning period, see § **3.05(d)**.

The current move to structuring Farmouts as *Tax Partnerships* has been discussed in § **3.04(b)[2][c-5-b]** and its subsections. Here, the provisions of a Joint Operating Agreement attached as an Exhibit to the Farmout Agreement can create unexpected problems for the parties. Since all versions of the AAPL Model Form Operating Agreement contained provisions under which the parties elect to be *excluded* from the provisions of Subchapter K of the Internal Revenue Code,[181] inconsistencies can exist between these Operating Agreement provisions and a provision in the Farmout Agreement electing *not* to be excluded from the provisions of Subchapter K.[182] While the Farmout Agreement will often contain a provision that its terms will control over any inconsistencies with attached exhibits, the contrary Operating Agreement election might

---

181 See Article IX of the 1982 and 1977 versions, and § 26 of the 1956 version of the AAPL Model Form.

182 As to the effects of Subchapter K Election, see § **3.04(b)[2][c-5-b-2]**.

be treated as terminating the Tax Partnership once Joint Operations are commenced.[183] (In some instances, a Farmee also finds that he has been "sandbagged," either intentionally or otherwise, by a provision in the Operating Agreement electing not to be excluded from the provisions of Subchapter K — thus forming a Tax Partnership — when the parties had neither discussed, nor did the Farmee intend, that the Partnership be formed.)

Other problems that may arise under the Joint Operating Agreement are discussed in detail in **Chapter Four**.

A number of other matters affecting the conduct of the Farmout operations are frequently covered in Exhibits. Thus, the Farmout Agreement may contain Exhibits describing in detail the leases which are the subject of the Farmout (see § § **3.01(b)[2]** and **3.02** *et seq.*); the details of the tests to be performed by the Farmee (see § § **2.04(b)** *et seq.* and **3.03(c)**); an accounting procedure specifying how the various "cost" provisions of the Farmout will be measured (see § **3.04(b)[3][b]** and **Chapter Four**); and an Insurance Schedule (see § **3.05(d)**), to mention only a few.

§ 3.05(d) Miscellaneous Provisions. There are a number of additional provisions that should be contained in the Farmout Agreement, or are desirable (or looked at as being desirable) to complete the contract between the parties. Certain of these provisions, such as "Confidentiality of Information" and Indemnification provisions have been discussed earlier in connection with the Support Agreement.[184] The Agreement will also normally contain a

---

[183] As to the consequences of such a premature termination, see § **3.04(b)[2][c-5-b-4]**.

[184] As to Confidentiality of Information provisions, see § **2.06(c)**. As to Indemnification, see § **2.06(b)**.

provision that "time is of the essence,"[185] as well as adequate Notice provisions.

In connection with the Farmout, the Farmor may need to designate several company representatives to be notified for various purposes, such as its Production Superintendent, separate Geological and Petroleum Engineering representatives, etc. If several persons are so designated, the Agreement should specify which person is to be notified for a given purpose, and who has the option to elect and/or give consent or approval for the company for various purposes.[186]

Just as Insurance provisions are required in Joint Operating Agreements,[187] the Farmor will normally wish to bolster the Indemnification provisions of the Farmout Agreement by requiring the Farmee to carry specified Insurance. The Agreement will normally give the Farmor a right to request delivery of Certificates of Insurance prior to the commencement of operations, evidencing the specified coverage in companies acceptable to the Farmor, and requiring notification in the event that the Insurance is canceled. (The insurance carrier should also be requested to similarly notify the Farmor upon such a cancellation.

The Farmee should also expressly be obligated to comply with the various terms of the Farmout Leases and any intermediate Assignments or other third-party obligations.

The Farmout Agreement will normally contain a provision under which the Farmee's rights will be non-assignable during the earning period, at least without the consent of the Farmor. If the Farmee

---

[185] In the absence of such a provision, the general rule is that dates for performance are *approximations only* and need not be strictly complied with. However, some courts have stated that time is always of the essence in oil and gas transactions.

To avoid any question, an express "time is of the essence" provision is desirable.

[186] Questions concerning *Notice* provisions in general are discussed in greater detail in **Chapter Four**.

[187] See **Chapter Four**.

knows that it will be necessary for him to "layoff" a portion of his interest to other parties, he should request that the Agreement specifically permit assignment to such parties, or, at a minimum, that the Farmor's consent to assignment will not be "unreasonably" withheld. Likewise, subject to any negotiated "preferential right to purchase" provision (see § **3.04(c)**), there should be no restrictions on the Farmee's right to assign once the earning conditions have been satisfied.

The importance of defining contract terminology has been mentioned previously in connection with well commencement and completion provisions, and rights assigned and reserved (see § § **3.03(b)[1]** and **3.04(b)[3][b]**), and will be discussed in greater detail in **Chapter Eight**. Again, use of concise and precisely-drafted "defined terms" can add both to the clarity and brevity of the Agreement.

The Agreement should finally provide a "time for acceptance," to insure that a prospective Farmee has been negotiating in good faith and is not merely using the Farmout proposal as an opportunity to haphazardly "shop" the deal or — even worse from the Farmor's standpoint — to "Top Lease" the Farmor's Leases.[188]

Certain of the provisions discussed in this section may be looked at as mere "boilerplate." Certainly, provisions such as the Notice, Indemnification, and "time is of the essence" terms are found in the majority of oil and gas contracts. Likewise, a number of optional provisions may be sprinkled throughout the Agreement as the parties see fit. These would include such clauses as a *Captions* provision, specifying that the heading of a section will have no effect or limitation on the interpretation of the Agreement, or a provision specifying that references to parties in the neuter will also include the masculine and the feminine and that references in the singular will include the plural.

---

188 As to "Anti-Top-Leasing" protections, see § **3.01(b)[6]**.

The term "boilerplate" suggests that the provisions are of no real importance to the parties and can be skimmed over. Tell this to someone who has found that his Agreement lacks an adequate Notice provision when Notice — and how it is to be implemented — has suddenly become critical to the deal! Making sure that these "miscellaneous" clauses are properly drafted, and that all necessary "boilerplate" has been included in the Agreement, can thus be just as important to the parties as adequate provisions in the "key" negotiation areas.

INDEX

TOPICAL INDEX TO VOLUME ONE*

* Author's Note: A detailed Index to Volumes One through Four is contained in Volume Five.

D

D

E

F

G

I

J

L

M

P

Q

R

S

T

U

W

Y